T0310648

SUSTAINABLE LUBRICATION

SUSTAINABLE LUBRICATION

Jitendra Kumar Katiyar, Ranjeet Kumar Sahu, and T. C. S. M. Gupta

CRC Press is an imprint of the
Taylor & Francis Group, an **informa** business

First edition published 2022
by CRC Press
6000 Broken Sound Parkway NW, Suite 300, Boca Raton, FL 33487-2742

and by CRC Press
2 Park Square, Milton Park, Abingdon, Oxon, OX14 4RN

CRC Press is an imprint of Taylor & Francis Group, LLC

Library of Congress Cataloging-in-Publication Data
Names: Katiyar, Jitendra Kumar, author. | Sahu, Ranjeet Kumar, author. | Gupta, T. C. S. M., author.
Title: Sustainable lubrication / Jitendra Kumar Katiyar, Ranjeet Kumar Sahu, T C S M Gupta.
Description: First edition. | Boca Raton, FL : CRC Press, 2022. | Includes bibliographical references and index.
Identifiers: LCCN 2021055081 (print) | LCCN 2021055082 (ebook) | ISBN 9781032061962 (hbk) | ISBN 9781032062068 (pbk) | ISBN 9781003201199 (ebk)
Subjects: LCSH: Lubrication and lubricants. | Green chemistry.
Classification: LCC TJ1075 .K285 2022 (print) | LCC TJ1075 (ebook) | DDC 621.8/9--dc23/eng/20220103
LC record available at https://lccn.loc.gov/2021055081
LC ebook record available at https://lccn.loc.gov/2021055082

ISBN: 978-1-032-06196-2 (hbk)
ISBN: 978-1-032-06206-8 (pbk)
ISBN: 978-1-003-20119-9 (ebk)

DOI: 10.1201/9781003201199

Typeset in Times
by MPS Limited, Dehradun

Contents

Preface

Over the past few decades, due to the stringent regulations on emission standards, strict exhaust emission guidelines were imposed on industries to protect and save the environment and excessive depletion of energy resources. To overcome the problems, the quest for alternate raw materials, which are sustainable/renewable to replace fossil fuel-based material, has become a focus of research attention. Most of the lubricants used in industries are based on mineral oils and formulated with required additives for a variety of lubrication purposes. The purposes include minimal frictional heating, minimal wear, and enhanced component life by enhancing the service life. Presently, the global consumption of lubricants is approximately 37.4 million tons, of which the automotive sector consumes the majority of lubricants (almost 68%) and the rest 32% is consumed by other industries. This 32% of lubricants include 12% hydraulic oils, 15% metalworking and cutting fluid, 3% greases, and 2% gear oil. The use of these lubricants and their indiscriminate disposal subsequently produce environmental problems such as pollution in water and soil, and further, the degradation of raw materials in the environment is of serious concern. Therefore, alternate synthetic lubricants such as polyalphaolefin, polyalkylene glycol, and esters were developed, and these lubricants were found to either extend the life of the lubricant (drain interval) or dispose of their byproducts safely after the end of service life. They are formulated by chemical synthesis rather than the refinement of existing petroleum or vegetable oils. They are generally superior to the mineral oil lubricants in terms of thermal-oxidative responses, and therefore extend their service life. Thus, the consumption of synthetic lubricants is increasing. However, the high price of synthetic lubricants impacts the market growth. Keeping this factor in view, industrial researchers have focused more on biodegradable base oils (i.e. green lubricants) which were derived from plant oils and animal fats. The investigation of the development of bio-lubricants has received significant attention due to the fact that 50% of all lubricants worldwide will end up in the environment through spillage, usage, or improper disposal. Compared to petroleum-based lubricants the bio-lubricants extracted from renewable origin show higher lubricity, lower volatility, higher shear stability, higher viscosity index, and higher load-carrying capacity. Apart from these advantages of bio-lubricants, the biggest drawbacks include poor thermal/oxidative stability, high pour points, and inconsistent composition. Due to these drawbacks, researchers have tried to develop chemically modified bio-lubricants or to use suitable additives in them. The additives could improve the lubricant performance of the base fluid. Without additives, even the best base fluids are deficient in some features.

Anti-wear and extreme pressure additives have become the focus of considerable research atiention in recent decades. These additives include different carbon-based materials at nano- or micro-level such as graphite, graphene, multi/single-walled carbon nanotubes, fullerenes, etc., and others are hexagonal boron nitride, molybdenum disulfide, phosphorus, sulfur, zinc dialkyldithiophosphates (ZDDP), and many more. Mostly, such types of lubricants may show an excellent performance compared to synthetic or mineral-based lubricants. The field of such type of bio-lubricant also covers the fully or partially recyclable oil and its mixture. These lubricants should also be more cost-effective than synthetic oils. However, the optimization of cost is not an easy task. More research is required for the development of bio-lubricant properties to meet the rising demand in today's market. Further, the application of lubrication is also changing day by day for macro- to nano-level components because of the change in tribological behavior of the components.

Hence, this book describes the recent developments in sustainable lubricants and their usage in tribological systems. It also describes the new kind of base fluids. Further, it explains the optimization of a new type of lubrication according to the tribological systems. These tribological systems may be gears, bearings, micro-electro-mechanical systems, production equipment such as metal forming, sheet metals, metal cuttings, etc. In addition, few tribo-systems worked at very high temperatures such as turbines, which require high-temperature lubricants for enhancing the performance of equipment and service life. Moreover, where bulk lubrication cannot be used such as in micro/nano level components, the surface is protected either by self-assembled monolayers or by *in-situ* lubrication. Therefore, the optimization of lubricant properties widely depends on materials, geometry, applied load, working speed, working temperature, applied pressure, etc. The advances in lubricants' usages in various types of tribological systems are also discussed in this book.

Authors

Dr. Jitendra Kumar Katiyar is a Research Assistant Professor with the Department of Mechanical Engineering, SRM Institute of Science and Technology, Kattankulathur, Chennai, India. His research interests include tribology of carbon materials, polymer composites, self-lubricating polymers, lubrication tribology, modern manufacturing techniques, and coatings for advanced technologies. He earned his bachelor's degree from UPTU Lucknow with Honors in 2007. He earned his master's from the Indian Institute of Technology Kanpur, India in 2010 and PhD from the same institution in 2017. He is a life member of Tribology Society of India, Malaysian Society of Tribology, Institute of Engineers, India, The Indian Society for Technical Education (ISTE), etc. He has authored/co-authored/published more than 30 articles in reputed journals, 30+ articles in international/national conferences, 15+ book chapters, and published 6 books including as *Automotive Tribology* and *Tribology in Materials and Applications* with Springer, *Tribology and Sustainability*, *Green Tribology: Emerging Technologies and Applications*, *Biotribology: Emerging Technologies and Applications* with CRC Press/Taylor & Francis Group, USA and *Engineering Thermodynamics* for UG level with Khanna Publication. He has served as a member of the Editorial Board for *Tribology Materials, Surfaces and Interfaces* and Review Editor for *Frontiers in Mechanical Engineering: Tribology*. Further, he has served as a guest associate editor for special issues in *Tribology Materials, Surfaces and Interfaces*, *Journal of Engineering Tribology Part J*, *Arabian Journal for Science and Engineering*, *Industrial Lubrication and Tribology*, *Journal of Process Mechanical Engineering* and *Frontiers of Mechanical Engineering*. He is also an active reviewer in various reputed journals related to materials and tribology. He has delivered more than 30 invited talks on various research fields related to tribology, composite materials, surface engineering, and machining.

Dr. Ranjeet Kumar Sahu earned his BE degree in mechanical engineering 2002 from Berhampur University, Odishia, India, M.Tech degree in Production Engineering in 2011 from NIT Rourkela, Odisha, India, and PhD in nano-manufacturing from Indian Institute of Technology Madras, Tamil Nadu, India in 2016. Currently, he is an Assistant Professor with the Department of Mechanical Engineering, National Institute of Technology Karnataka, Surathkal, India. He received Institute Day Best Scholar Award in 2010, Best M.Tech Award in 2012, and Prof. M. S. Shanmugam Best Ph.D. Thesis Award in 2016. He has published many papers, books, and book chapters at the national and international level. His current research areas of interest include micro/nano-machining, nano-materials synthesis and characterization, precision engineering, and additive manufacturing.

Dr. T. C. S. M. Gupta is a Senior Vice President, R&D, Quality and Technical, Apar Industries Limited Navi Mumbai, India. He worked under CSIR scholarship at the Indian Institute of Petroleum, Uttarkhand, India for his PhD (HNB Garhwal University, Uttarkhand, India) and master's in chemistry from the University of Roorkee. He has co-authored/published 50+ papers, 2 Indian patents, and 6 commissioned technical feasibility reports. He is a member/working group lead in professional societies such as Cigre France, IEC, BIS, Tribology Society of India, Malaysia. Dr. Gupta is currently Secretary of NLGI India and Vice Chairman of Association of Lubricants Manufacturers Union (ALMU), Singapore. He is an approved supervisor/examiner/advisory panel member of Nottingham University, Taylors University, Malaysia ICT Mumbai, Maharashtra, India, PDPU Gandhinagar DDU, Gujarat, India, Nadiad, Gujarat, India for master's and doctoral programs. He worked as Chief Technical and Operating Officer for an independent lubricant company in Malaysia and traveled extensively for professional and technical conferences across the globe.

Introduction to Lubrication and Lubricant Additives

1

INTRODUCTION

Lubrication has been used by mankind from thousands of years. Water and oil are the oldest lubricant which was found by humans. Even human body has natural lubrication at every bone joints that is known as synovial fluid. Lubrication helps to reduce wear and heat that is produced by friction. Now a days machine are used in extreme working condition such as varying loads, speed, and temperature, etc. To reduce the friction and wear, the effective lubrication is required in such a specific application that can be achieved by addition of the suitable additives. Lubricant additives are added to the base oil for enhance the properties of the lubricating system as well as to impart the new properties of the lubricant, which helps the lubricant to use at various environment conditions. The development of lubricant additives may leads to the extended tool life, increase efficiency of the components, improving machining operation. Further, the different types of additives have also shown their unique properties that has been discuss in this chapter.

1.1 LUBRICATION

Friction leads to the formation of wear and heat, which can be minimized by the addition of lubricant between the mating surfaces. Therefore, lubrication is

DOI: 10.1201/9781003201199-1

applied between two mating surfaces that are moving in a relative motion to minimize the wear and heat followed by improving the efficiency of the component. The performance of a lubricant depends on its chemical, physical, and rheological properties. It also depends on the external parameters imposed such as contact pressure, relative speed, and temperature. Further, it helps to reduce oxidation, prevent rust, and also remove impurities/debris (Bartz, 1974).

1.1.1 Lubrication Regimes

Reynolds discovered that when lubrication was supplied to the shaft, it pushed a converging wedge of lubricant between the shaft and the bearing. He also noticed that as the shaft accelerated, the liquid flowed faster between the two mating surfaces. This is due to the viscosity of the lubricant that creates enough liquid pressure in the lubricant wedge to keep the two mating surfaces apart. Further, the liquid lubrication can be studied by a plot known as the "Stribeck curve" or "Stribeck–Hersey curve". The "Stribeck curve" or "Stribeck–Hersey curve" (named after Richard Stribeck and Mayo D. Hersey; shown in Figure 1.1) was developed in the first half of the 20th century to categorize the friction properties between two liquid lubricated surfaces. The curve is plotted with a friction coefficient as a function of a parameter given by $\eta U/P$; where η is the viscosity of the lubricant, U is the relative velocity, and P is the contact pressure.

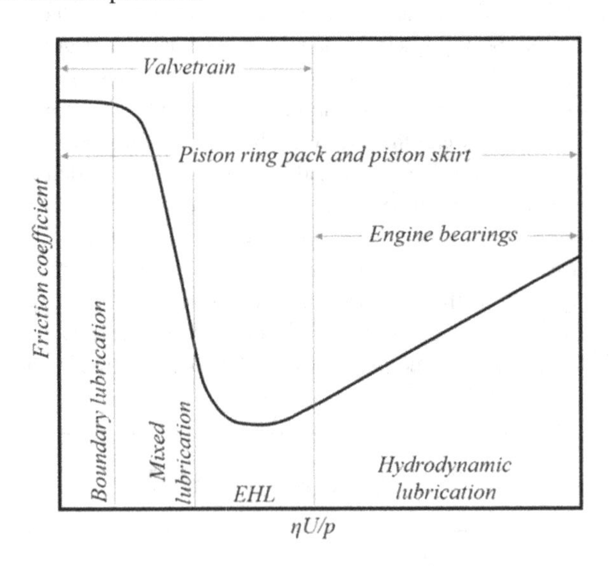

FIGURE 1.1 Stribeck curve for engine components (Tung & McMillan, 2004).

The Stribeck curve is basically a curve between coefficient of friction and bearing number. It shows the different regimes of lubrication. The lubrication regimes are broadly classified into four regimes: boundary lubrication, mixed lubrication, elastohydrodynamic lubrication (EHL), and hydrodynamic lubrication. In boundary lubrication condition only the asperities carry the load, thus friction between the moving surfaces is high. It is also called a solid lubricant. In hydrodynamic lubrication, the relative velocity is essential and lubricant viscosity plays a major role when compared to the boundary lubrication. In hydrodynamic or fluid film lubrication there is no wear because surfaces are not in contact. When the coefficient of friction is the minimum often, it is known as EHL, and lies in the interface of mixed lubrication and hydrodynamic lubrication. Mixed lubrication is the regime where the full hydrodynamic or EHL and boundary lubrication takes place. It is also called partial lubrication. Both the EHL and metal-to-metal contact occurr in mixed lubrication, which means the load applied on the component acts on both fluid films and by surface asperities. Piston rings, cams, and engine bearings are the well-known examples of mixed lubrication (Delprete & Razavykia, 2020).

1.2 TYPES OF OIL LUBRICATION

1.2.1 Oil Bath Lubrication

Oil bath lubrication is widely used with low or medium speeds. It is the simplest lubrication method for pumps (Fatih Dokme, 2016). The oil level should be at the center of lowest rolling element. It is desirable to provide a sight gauge due to which the proper oil level may be maintained.

1.2.2 Drip-feed Lubrication

Drip-feed lubrication is commonly used to supply the required amount of lubricant at required intervals. It is used for low load and low-to-moderate speed components. In old days, it was predominantly performed by hands, which leads to high risk: later it was replaced by automated components. Drip-feed systems are also known as gravity feed systems. A drip-feed reservoir can be connected to only one bearing, known as a single-point system, or to several bearings, known as a multiple-point system, via pipes or ducts. However, a pressurized drip-feed system can also be used to provide oil for a bearing system. The reservoir is partially or totally transparent, and when reaches a particular level, it is refilled manually.

1.2.3 Splash Lubrication

In the splash lubrication system of the engine, oil is splashed onto cylinder walls for each revolution of the crankshaft. Engine walls, piston rings, crankshaft bearings, and large-end bearings are all affected by this motion. The splash system mostly works in the connection with the pressure system in an engine; some parts are lubricated by splash system, whereas the other parts by a pressure system.

1.2.4 Circulating Lubrication

Circulating lubrication is commonly used for a high-speed operation at elevated temperatures. Oil is supplied to the system at one end through the pipe, travels through the component to lubricate, and drains out through the pipe at other side and then is stored in the reservoir for next circulation after filtration.

1.2.5 Oil/Air Lubrication

In this type of lubrication system, the lubricating oil with air premixed via nozzle is sprayed on the tool surface during machining operation. It is also called minimum quantity lubrication. This lubrication method is explained in detail in Chapter 4.

1.3 CLASSIFICATION OF LUBRICANTS

Lubricants are classified into three types based on their physical state: liquid lubricant, semi-solid or grease lubricant, and solid lubricant.

1.3.1 Liquid Lubricant

Liquid lubricants are also known as lubricating oils. Lubricating oils are further classified into three categories:

i. Animal oils and vegetable oils,
ii. Mineral or petroleum oils,
iii. Blended oils.

1.3.1.1 *Animal and Vegetable Oils*

Animal oils extracted from animal fats and vegetable oils are caster, coconut cotton seed, etc. These oils can easily adhere to the surface even under high temperatures and loads. But these oils undergo oxidation as compared to other oil. Therefore, they are rarely used for lubrication these days or if used then oxidation preventing agent is added to the oil.

1.3.1.2 *Mineral or Petroleum Oils*

These oils are available in abundance on earth. They are cheap and can be used in severe conditions. Mineral oil has low viscosity due to low molecular hydrocarbons (12 to 50 carbon atoms). To improve the viscosity, additives are added to mineral oil. These oils are distilled from the crude oil and collected at various temperatures.

1.3.1.3 *Blended Oils*

The desired properties cannot be achieved using a single oil. Thus, two or more base oils are blended along with the required additives to form a blended oil.

1.3.2 Semi-solid or Grease Lubricant

Grease is a semi-solid lubricant made by blending lubricating oil with thickening agents. The grease lubricant's main component is base oil, which can be either based on petroleum or synthetic oil. The thickener consists of special soaps such as Li, Na, Ca, Ba, Al, etc. Non-soap thickeners include carbon black, silica gel, etc. In Chapter 4, these are discussed in depth.

1.3.3 Solid Lubricant

Solid lubricant can be used in cases where the lubricating film did not perform its job by staying for a sufficient time. The solid lubricant is commonly used at higher temperatures and heavy loads. A solid lubricant can be in the form of powders, films, or composite materials. Some widely used solid lubricants are molybdenum disulfide, graphite, tungsten disulfide, zinc oxide, etc. These solid lubricants are highly anisotropic, with weak bonding between particular crystal planes and molecules. Thus, they possess self-lubrication properties, resulting in a low coefficient of friction. Graphite is the stacks of graphene held together by weak Van der Waals forces, so the energy required to shear the crystals are low. These parallel layers, which can easily slide one over the other, make graphite an effective lubricant. It is used either in powder form or as a suspension in oil or grease. It is soapy to touch; non-inflammable and stable up to a temperature of 375 °C. Further, molybdenum disulfide has a sandwich-like structure with a layer of molybdenum atoms in between two layers of sulfur atoms. Its poor interlaminar attraction helps these layers to slide over one another easily. It is stable up to a temperature of 400 °C.

1.4 PROPERTIES AND APPLICATIONS OF LUBRICATION

Lubricants have a wide range of physical and chemical properties that influence their performance. It is crucial to understand these properties when deciding which lubricant is suitable for required application. There are many properties of lubricants, but the most important properties are discussed next.

1.4.1 Viscosity

Viscosity is a measurement of a fluid's internal friction. It is the most important physical property of a fluid in the context of lubrication. A lubricant's viscosity is affected by temperature, pressure, and, in some situations, the rate at which it is sheared. The force is required by a fluid to overcome internal molecular friction and flow is known as dynamic viscosity. Dynamic viscosity is involved in the tribological analysis. The kinematic viscosity of a lubricant is the ratio of the viscosity of the fluid to the fluid's density (M. J. Neale, 2001). The most significant attribute of a lubricant is viscosity, which helps to form a lubricating layer that cools the machine components, seal, and manage oil consumption.

1.4.2 Viscosity Index

The viscosity index (VI) of a lubricant is the rate of the viscosity change due to a temperature change.

1.4.3 Oxidation Stability

When oxygen is combined with lubricating oil, a phenomenon is called oxidation. Further, ability of lubricant to resist the oxidation is known as oxidation stability.

The rate of oxidation will be accelerated by factors such as high temperatures, water, metal catalysts, and acids. As temperatures increases, the life of a lubricant decreases, resulting in the formation of varnish and sludge. The oxidation in the lubrication is affected the anti-wear performance (Eissa et al., 2010).

1.4.4 Pour Point

The pour point of lubricant is the temperature at which the rheological property of the oil sharply changes from liquid to semi-solid, in other words, the lowest temperature at which oil can flow. For example, the amount of the wax present in the lubricant will affect the gelation (Gavlin et al., 1953).

1.4.5 Demulsibility

Demulsibility of oil, or its ability to release water, is a very important characteristic of lubricating. Emulsified water causes the oil to appear hazy or milky, and it can also result in foam formation. Worse, the emulsified water reduces the oil's ability to provide lubrication; raises temperatures; and promotes corrosion, oxidation, and degradation of the oil.

1.4.6 Flash and Fire Point

The flash point is the lowest temperature at which the lubricant starts to sprinkle when an external ignition is provided. Similarly, the fire point is the lowest temperature at which the lubricant catches fire and burns continuously even after removing the ignition source (Thangarasu & Anand, 2019).

1.4.7 Density

It is the substance mass per unit volume. It is an important parameter for fluids.

1.4.8 Other Properties

The other properties include tendency to form foam and stability of the foam, detergency, dispersancy, seal compatibility thermal conductivity, surface tension, electrical resistivity, and dielectric constant. Some of the above-mentioned properties can be achieved only by the addition of suitable additives. Table 1.1 presents the lubrication properties tested at various ASTM standards.

TABLE 1.1 Lubricant properties and its standard test methods

PARAMETER	*STANDARD TEST METHOD*	*STANDARD*
Dynamic viscosity and kinematic viscosity	Dynamic viscosity and density of liquids by Stabinger Viscometer (and the calculation of kinematic viscosity)	ASTM D7042
Oxidation stability	Determination of the oxidation of used lubricants by FT-IR using peak area increase calculation	ASTM D7214
Pour point	Pour point of petroleum products	ASTM D97
Demulsibility	Water separability of petroleum oils and synthetic fluids	ASTM D1401
Flash and fire point	Flash point by Pensky-Martens closed cup tester/fire point by Cleveland open cup tester	ASTM D93 and ASTM D92

1.5 FUNCTION OF LUBRICANTS

- Reduce the wear and tear of the surfaces by avoiding direct metal-to-metal contact between the rubbing surfaces.
- Reduce the expansion of metal due to frictional heat and destruction of material.
- Act as a coolant e.g., engine oil.
- Avoid unsmooth relative motion.
- Reduce the maintenance cost.
- Reduce the power loss in internal combustion engines.
- Prevent the components from foreign material penetration, rust, and corrosion.

1.6 ADDITIVES

Additives are organic or inorganic chemical compounds dissolved or suspended as solids or liquid in base oil. They typically range between 0.1% and 30% of the base oil, depending on the requirement. Additives added to the base oil to alter the properties can be classified into three major roles: (i) to enhance existing properties, (ii) to suppress undesired properties, and (iii) to impart new properties. Additives enhance the existing properties such as anti-oxidants, corrosion inhibitors, anti-foam, and demulsifying characteristics; suppress undesired properties such pour point and VI; and impart new properties such as extreme pressure (EP), detergency, and dispersion.

1.6.1 Types of Additives

1.6.1.1 Pour Point Depressants Additive

Pour point depressants are additives that lower the lubricant's pour point, allowing it to remain liquid and maintain its fluidity (pourability) at lower temperatures. When the temperature of the base oil drops below 50 °C, paraffin molecules begin to crystallize as wax, and the oil loses its capacity to flow by gravity or be pumped under pressure. This impacts the oil's viscosity. Alkylaromatic polymers and polymethacrylates, for example, limit wax crystal development by changing the interface between the wax and the oil molecules, reducing the pour point by roughly 20–30 °F (11–17 °C) (Puhan, 2016). These additives help lubricant to work under extremely low temperature. Further, the synthetic oils along with the minerals oil with some of these additives can even improve the pour point temperature (Alaboodi, 2020).

1.6.1.2 Viscosity Index Improver Additives

Naturally, fluid viscosity is reduced when it is exposed to higher temperature. VI improvers are those additives that prevent the base oil from losing its viscosity at higher temperature. An ester-type chemical compound polyalkylmethacrylate (PAMA) is the first vico-situ improver additive used in the base oil. Because of this property, the same engine oil has been used in both summer and winter. Earlier, the VI additives were used in transmission, gear, and hydraulic fluids. During 80s, olefin copolymer (OCP) hydrocarbon base VI additives replaced the PAMA. PAMA has shown the best property as additives in base oil, such as a strong thickening effect at high temperatures and a much weaker thickening impact at low temperatures (Lauterwasser et al., 2016). In heavy-duty application, viscosity improver additives compress and fragment into small pieces due to the large compressive pressure acting between two mating surfaces. This elongation and compression of the polymer chain reduces the thickness of the lubricating film, but when high shear condition is removed the polymer chain returns to its original position. This phenomenon is referred as temporary shear thinning. Under extremely high pressure, the polymer chain gets fragmented into smaller chains that cannot be joined even after removing the load which is known as the permanent shear-thinning.

1.6.1.3 Anti-wear Agent

Anti-wear agent are restrained the two-body wear phenomenon at metallic counter surfaces in the boundary lubrication regime. In this regime, a thin film thickness is developed due to which the asperity-asperity contact exists. These additives are polar in nature, allowing them to bind with metallic surfaces and produce an anti-wear coating by tribochemical or mechanochemical interactions. Fatty oils, acids and esters are mixed with lubricating oil to reduce the friction at counter surface. Among all additives, Zinc dialkyldithiophosphate (ZDDP) is the most commonly used anti wear agent between 1930 and 1940. Further, ZDDP also provides antioxidant, EP additive and corrosion-inhibition properties to the lubricant. Moreover, ZDDP reduces the mild wear by preventing the metal surface from contacting each other, resulting in adhesive wear reduction. Due to its multi functionality, ZDDP is considered as the most cost-effective antioxidant, Extreme pressure and Antiwear additive among all available additives (Spikes, 2004).

1.6.1.4 Anti-oxidants

Anti-oxidants are also known as oxidation inhibitors that prevent the oxidation of the component and the base oil, which leads to an increase in lubricant life. Oxidation is high at higher temperature. The oxidation can be induced by water, wear particles, and other contaminants in the lubrication, which leads to the formation of acids and sludges. These formed acids initiated the corrosion on the metallic component, whereas the sludge formation increases the viscosity of the lubricant. The common anti-oxidant used in lubricating oil and grease lubrication is ZDDP, hindered phenol, sulfurized phenols, and aromatic amines.

1.6.1.5 Defoamants

Defoamants, also known as anti-foaming agents, are additives that prevent the lubricant from foaming. Foaming occurs due to the frequent mixing of oils with air or other gases, resulting in air entrapment. Foam prevents the items from cooling since it is a poor heat conductor. It lowers the load-carrying capacity and lubrication flow, resulting in increased engine wear. Silicone polymers, such as polymethylsiloxane as few PPM, and organic copolymers, such as alkoxy aliphatic acids, polyalkoxyamines, polyethylene glycols, and branched polyvinyl ethers, at higher concentration are widely used in mineral oils as the defoaming agents.

1.6.1.6 Friction Modifier

Friction modifiers are the lubricant additives commonly used in boundary and/or mixed lubrication conditions to adjust friction characteristics and improve the lubricity and energy efficiency. Friction modifiers are used in engine oil and automated transmission oil to alter the coefficient of friction between the sliding parts. There are two main types of friction modifiers for liquid lubricants: organomolybdenum compounds and organic friction modifiers. The former friction modifiers can be divided into three families: sulfur- and phosphorus-containing compounds, such as molybdenum dialkyldithiophosphates (MoDTP); sulfurcontaining and phosphorus-free compounds, such as molybdenum dithiocarbamates (MoDTC); and sulfur- and phosphorus-free compounds, such as molybdate ester. Generally, long-chain surfactants with polar end groups, including carboxylic acid, ester, alcohol, amine, amide, imide, borate, phosphate, ionic liquid, and their derivatives are used as friction modifiers. They preferentially adsorb very strongly onto the metallic surface. The chemical structure and the polarity of the molecules play a major role in the friction reduction (Tang & Li, 2014).

1.6.1.7 Detergents and Dispersants

Lubricant detergents are metal salts of organic surfactants that provide corrosion protection, deposit preventions, and other performance enhancement. Sulfonate-, phenate- and salicylate-type detergents incorporating calcium carbonate are the most widely used lubricant detergents. Further, the dispersants are used mainly in engine oil along with detergents to keep engines surfaces clean and free from deposits. Dispersants keep the insoluble soot particles that are produced during combustion in the IC engine that are finely dispersed or suspended in the lubricant even at high temperatures. These suspended particles are periodically removed through oil filtration or oil change. Thus, dispersants minimize the damage to engine components such as piston, cylinder wall, etc. Polymeric alkylthiophosphonates, alkylsuccinimides, succinic acid esters/amides, and their borated derivatives, as well as organic complexes containing nitrogen compounds, are used as the dispersants.

1.6.1.8 Corrosion and Rust Inhibitor

Corrosion and rust inhibitors are additives that reduce or eliminate the rust formation on the component by neutralizing acids and forming a protective film on the metal surface. Due to the physical or chemical interaction of the lubricating film, the water reaching the metal surface is retarded. These are usually the compounds having a high polar attraction toward metal surfaces such as succinates, alkyl earth sulfonates, metal phenolates, fatty acids, amines as well as zincdithiophosphates. As per the requirement, several metal types of corrosion inhibitors are used. These additives form a physical barrier on the metal surface in the form of a dense hydrophobic monolayer of chemisorbed surfactant molecules, which prevents water and oxygen from reaching the metal surface.

1.6.1.9 Extreme Pressure (EP) Additives

When high temperatures and strong loads are applied to gear or bearing, EP additives are necessary to reduce friction, control wear, and avoid severe surface damage. They react chemically with metallic surfaces to generate a sacrificial surface film that prevents welding and seizure of asperities at the metal-to-metal contact. These additives are added in the formation of a smooth surface in asperities, allowing the load to be distributed uniformly throughout the surface, reducing wear, and resulting in effective lubrication. The effectiveness of EP additives is measured by their capacity to produce a thick surface coating under high load and at high contact temperatures. EP additives usually contain sulfur and phosphorus compounds and chlorine or boron compounds. Ashless EP additives such as dithiocarbamates, dithiophospates, thiolesters, phosphorothioates, thiadiazoles, aminephosphates, phosphites may be preferred in some applications where chlorine may cause corrosion. Therefore, depending on the area of application the suitable EP additives are used.

1.6.1.10 Nanoparticles as Lubricant Additives

Nanoparticles as lubricant additives have reduced the wear and friction in lubricant formulations while also improving metal-forming tool tribological performance. The addition of small amount of nanoparticles shows effective lubrication to the base stock, which results in significant improvements in the lubricant's properties. Tribological features, anti-oxidation capability, and thermal properties of lubricants can all be improved with the proper nano additives (Alaboodi, 2020). In recent years many nanoparticles including metals (e.g., Zn, Al, Cu, Ti, Fe) and their oxides, molybdenum disulfide and tungsten disulfide, metal borates, hexagonal boron nitride, fluorinated compounds fullerenes, graphitic nanoparticles, and nanodiamond particles are used as additives to enhance the properties of the liquid lubricant. Ivanov and Shenderova (2017) studied the dispersion of detonation nanodiamond (DND) particles in the polar solvent that is mixed with Mobil 5W30 lubricant at various concentrations and compared it with conventional soot dispersion as shown in Figure 1.2.

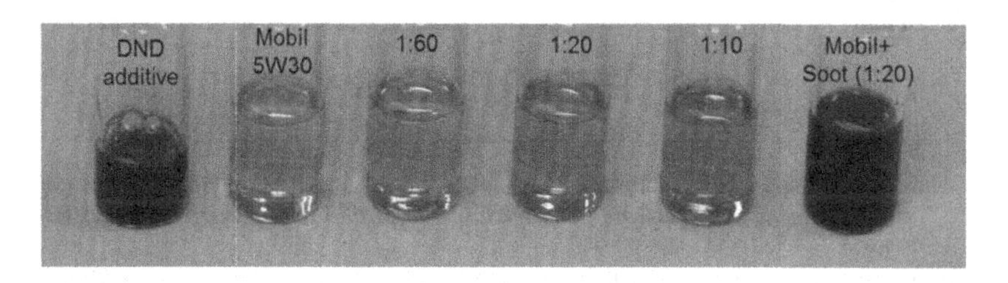

FIGURE 1.2 Photographs of the 20 nm DND-based additive (far left), pure Mobil 5W30 oil (next), and 3 vials of Mobil with DND additive at different concentrations, Mobil with a traditional soot additive is illustrated on the far-right image (Ivanov & Shenderova, 2017).

Due to the abrasive property of the nanodiamond particle, it can polish the asperities and decrease the friction, while reducing wear through the formation of a robust tribofilm. Further, Mohan Rastogi et al. (2021) studied the tribology properties of the 30 nm silica nanoparticles dispersion in jatropha oil without any surfactant. The addition of nanoparticles has also affected the viscosity and flash point of the base lubricating oil. The result showed the reduction in coefficient of friction by 0.6 wt% of nanoparticle addition. They also stated that increasing the nanoparticle may negatively affect the tribological properties. More studies are on going in dispersal of nanoparticle additives for the effectiveness to the base oil for enhancing the properties.

1.7 SUMMARY

Lubricants are vital for the tribological life of the machine elements. Nowadays the machine element is required to work under various environments in harsh conditions. The lubricant should stay longer for good maintenance and less cost. Lubricant additives are the only option to enhance the performance of the lubrication. It should be optimized to meet the required performance of the component. The lubricating additives should be environmentally friendly and last longer. Blending the different materials of additives shows synergistic effect that leads to the improved tribology properties. Nanoparticles are the future additive as they show immense performance in very less concentration.

REFERENCES

Alaboodi, A. S. (2020). Natural oils as green lubricants in forming processes. In: *Encyclopedia of Renewable and Sustainable Materials*. Elsevier Ltd. 10.1016/b978-0-12-803581-8.10849-5

Bartz, W. J. (1974). Lubrication and lubricants. *VDI-Z, 116*(12 August, 1974), 1013–1021. 10.1243/jile_proc_1921_011_030_02

Delprete, C., & Razavykia, A. (2020). Piston dynamics, lubrication and tribological performance evaluation: A review. *International Journal of Engine Research, 21*(5), 725–741. 10.1177/1468087418787610

Eissa, E. A., Basta, J. S., & Ibrahim, V. (2010). The oxidation stability of lubricating oil. *Petroleum Science and Technology, 28*(15), 1611–1619. 10.1080/10916460903160800

Fatih Dokme, M. T. (2016). Effects of Using Pure Eco-Friendly Lubricants in Pump Bearings Instead of Traditional Mineral Lubricants. International Conference on Environmental Science and Technology, 28th September to 2nd October, Belgrade Serbia, ISBN: 9786056626296.

Gavlin, G., Swire, E. A., & Jones, S. P. (1953). Pour point depression of lubricating oils. *Industrial & Engineering Chemistry, 45*(10), 2327–2335. 10.1021/ie50526a050

Ivanov, M., & Shenderova, O. (2017). Nanodiamond-based nanolubricants for motor oils. *Current Opinion in Solid State and Materials Science, 21*(1), 17–24. 10.1016/j.cossms.2016.07.003

Lauterwasser, F., Bartels, T., Smolenski, D., & Seemann, M. (2016). Megatrend Fuel Economy: How to Optimize Viscosity with VI Improvers. *SAE Technical Papers, 2016-February*(February). 10.4271/2016-28-0030

Mohan Rastogi, P., Kumar, R., & Kumar, N. (2021). Effect of SiO_2 nanoparticles on the tribological characteristics of jatropha oil. *Materials Today: Proceedings, 46*(xxxx), 10109–10112. 10.1016/j.matpr.2020.09.377

Neale, M.J. (2001). C-1 Viscosity of Lubricants, Lubrication and Reliability Handbook, Elsevier, Butterworth-Heinemann, 1–4, https://doi.org/10.1016/B978-075065154-7/50118-2.

Puhan, D. (2016). Lubricant and lubricant additive. In *Intech: Vol. I* (Issue tourism, p. 13). 10.5772/intechopen.93830

Spikes, H. (2004). The history and mechanisms of ZDDP. *Tribology Letters, 17*(3), 469–489. 10.1023/B:TRIL.0000044495.26882.b5

Tang, Z., & Li, S. (2014). A review of recent developments of friction modifiers for liquid lubricants (2007-present). *Current Opinion in Solid State and Materials Science, 18*(3), 119–139. 10.1016/j.cossms.2014.02.002

Thangarasu, V., & Anand, R. (2019). Physicochemical fuel properties and tribological behavior of *Aegle marmelos* correa biodiesel. In: *Advances in Eco-Fuels for a Sustainable Environment*. Elsevier Ltd. 10.1016/b978-0-08-102728-8.00011-5

Tung, S. C., & McMillan, M. L. (2004). Automotive tribology overview of current advances and challenges for the future. *Tribology International, 37*(7), 517–536.

Types of Lubricants

2

INTRODUCTION

In the contemporary industrial world, without high performance lubricants there would be much more wear, friction and corrosion leading to a tremendous waste of resources. The impact of the lubricant industry on helping to preserve the planet cannot be underestimated. However, the lubricant industry continues to improve their products and production in order to minimize the resources needed to manufacture high quality products. In recent years, green lubricants i.e. environmentally acceptable lubricants are increasing considerably in their use and has become an attractive and booming topic. The green lubricants will provide a sustainable future path for new promising applications. Therefore, in this Chapter, at the outset the broad classification of lubricants followed by the description of each type of usually used lubricants and their sub-types are detailed. Further, the conceptual realization of advanced lubricants such as ionic liquid, biodegradable and nano lubricants, and their importance in today's industrial society are addressed in the subsequent sections.

2.1 CLASSIFICATION OF LUBRICANTS

Lubricants reduce friction, resulting in a reduction of energy consumption and increased equipment/component life. The lubricants are used in various industrial, medical, and domestic applications. Traditionally, a lubricant contains a mixture of two main ingredients – oil and additives, for grease, a third ingredient – a thickener (Gow, 2009) and for paste, fourth ingredient solid

DOI: 10.1201/9781003201199-2

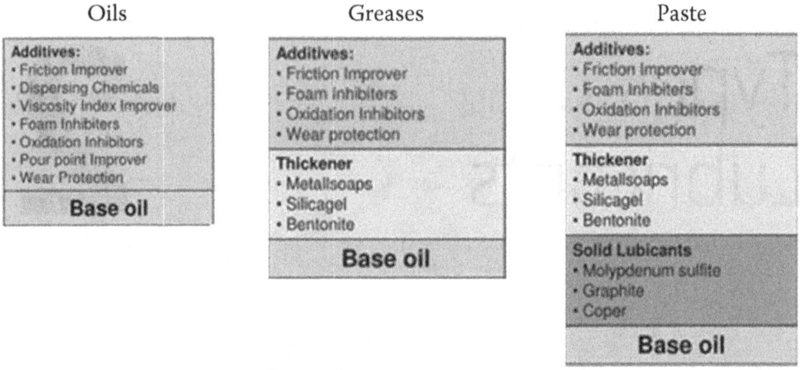

Oils

Additives:
• Friction Improver
• Dispersing Chemicals
• Viscosity Index Improver
• Foam Inhibiters
• Oxidation Inhibitors
• Pour point Improver
• Wear Protection

Base oil

Greases

Additives:
• Friction Improver
• Foam inhibiters
• Oxidation Inhibitors
• Wear protection

Thickener
• Metallsoaps
• Silicagel
• Bentonite

Base oil

Paste

Additives:
• Friction Improver
• Foam Inhibiters
• Oxidation Inhibitors
• Wear protection

Thickener
• Metallsoaps
• Silicagel
• Bentonite

Solid Lubicants
• Molypdenum sulfite
• Graphite
• Coper

Base oil

FIGURE 2.1 General types of lubricants.

lubricants are used. Figure 2.1 shows the general types of lubricants. In modern times, the lubricant formulation still follows this basic mixture, but the options have expanded dramatically, as many types of natural and synthetic base fluids can be used as the base of a lubricant, not just petroleum oil. Additives are included to impart beneficial performance attributes, such as reduced friction, corrosion protection, heat removal, foam and air release, water separation or emulsion, etc. (Habereder et al., 2008).

Based on the molecular structure of the lubricant material as well as its shear strength, lubricants are broadly classified as follows:

i. Solid lubricants
ii. Semi-solid lubricants
iii. Liquid lubricants
iv. Ionic liquid lubricants
v. Biodegradable lubricants
vi. Nano lubricants

2.2 SOLID LUBRICANTS

A solid lubricant is a solid material applied or inserted between two moving surfaces or bearing surfaces. This material will shear a lot more easily than the bearing or moving surfaces. The three main requirements for a material to be a solid lubricant are the ability to support the applied load without major distortion, a low coefficient of friction, and a low rate of wear. Solid lubricants are generally used when the conditions are extreme. Solid lubricants are used in powder form, as lubricating grease, in suspensions, in metallic films, or in bonded lubricants. Anti-friction linings use solid lubricants. The durability of solid lubricants is increased by coating the binders along with lubricating pigments. These bonded coatings provide greater film thickness and increase the wear life of the lubricant and the surface on which the lubricant is applied. The popular applications of bonded coating of solid lubricants are cylindrical brushes, separator cage of rolling bearing, and electrical brushes. Solid lubricants can be further classified into four subtypes, namely polymer, metal-solid, graphite, and ceramic and cermet.

2.2.1 Polymer Lubricants

Polymers are one of the largest groups of solid lubricants. They are suitable for use with light loads. They have a lower thermal conductivity which is the amount of heat that they are able to dissipate. There are three main polymer solid lubricants: polytetrafluoroethylene (PTFE), nylon, and synthetic polymers.

PTFE is a polymer derived from ethylene. All the hydrogen atoms in an ethylene molecule are replaced by fluorine atoms to give PTFE. It is more popularly known as Teflon, a trade name given to PTFE by the famous company Du Pont. It is commonly used as a solid lubricant because of its low friction, chemical stability, low surface energy, and greater chemical inertness. PTFE is also nontoxic and hence suitable for use in industries such as food and pharmaceuticals. For all the positive points of PTFE, there are some downsides as well. First, it has a relatively high rate of wear. Second, it has high thermal expansion and low thermal conductivity which makes it less desirable for use in high-temperature environments. Lastly, it has a low load capacity. However, many of these disadvantages can be addressed through the use of synthetic polymers. Synthetic polymer lubricant can be prepared by mixing glass and carbon-based fillers with PTFE. Impregnating PTFE with metal structures such as bronze or lead is also an option. This modification also allows the synthetic PTFE to withstand higher loads and improve the wear rate.

2.2.2 Metal Solid Lubricants

These solid lubricants contain lamellar solids and achieve low friction through a process known as film transfer. Molybdenum disulfide is the most commonly used metal-solid lubricant. Some of the advantages of a metal-solid lubricant such as molybdenum disulfide are high load-carrying capacity, good high-temperature performance, and low friction. It is also stable in vacuum up to 1000 °C. Hence, molybdenum disulfide also finds use in space applications. Its disadvantages are sub-optimal performance in the presence of moisture and high film thickness. A thicker film does not last as long because it is more prone to wear and tear.

2.2.3 Graphite Lubricants

Graphite seals are used as solid lubricants. This lubricant exhibits desirable properties such as high-temperature stability, high oxidation stability, and sustainable performance in high sliding speed applications. Graphite as a material has low friction and can withstand moderate loads. However, it is prone to corrosion and does not work very well in a vacuum. The lubrication performance of graphite actually increases with an increase in temperature. However, beyond 500 °C, the incidence of corrosion increases.

2.2.4 Ceramic and Cermet Lubricants

Ceramic and cermet coatings are used as lubricants in situations where a lower wear rate is more important than low friction. Ceramic/cermet coatings can be used at high-temperature ranges of around 1000 °C. A 0.5-mm-thick coating of ceramic/cermet material offers a low-cost way of utilizing its wear resistance. The coating can be sprayed using a detonation gun, plasma spraying, or electrolytic deposition using an electrolyte that contains ceramic particles.

2.3 SEMI-SOLID LUBRICANTS

Greases are also known as semi-solid lubricants. It can be considered one of the most versatile forms of lubricants. It can be used in a wide range of environments across various temperatures, load conditions, or speeds. Be it a dry or wet environment, dusty or clean environment, or even corrosive environment, grease finds use in all sorts of applications.

Grease is a type of pseudo-plastic fluid. One of the most significant properties of grease is consistency. Consistency is nothing but the relative hardness or softness of any material. Grease consists of lubricating oils that have low viscosity and are thickened by finely dispersed solids known as thickeners. Grease is made up of a combination of base oil, additives, and thickener.

Petroleum and synthetic base oils are used in the manufacturing of grease. The properties of the base oil are very important, as they affect the properties of the grease produced from the oil. Low viscosity and light base oil are used to produce grease that works at low temperatures. High-viscosity base oil is used to produce high-temperature grease. Certain chemical additives are added to the grease in order to improve its properties. The choice of additives depends completely on the end-use or application of the grease. Factors such as performance parameters, environmental impact, sustainability parameters, compatibility, cost, and color all play a role in the choice of additives. Thickeners are added to the base oil in order to thicken the material and produce grease. There are two types of thickeners, organic thickeners and inorganic thickeners. Inorganic thickeners are non-soap-based while organic thickeners are usually soap-based.

The advantages of using grease as a lubricating material include water resistance, binding strength to the surface, lower frequency of application, ability to reduce noise and vibration, ability to seal against contaminants, and usability with vertical/inclined shafts. The disadvantages include poor heat dissipation, vulnerability to being contaminated by dust, and the inability to filter out contaminants from the grease.

Lubricants may also be used in a paste form in heavy load applications, in sliding applications, and with slow-running bearings. The paste form of lubricant may also be used as assembly paste or as a high-temperature paste.

2.4 LIQUID LUBRICANTS

Liquid lubricants are used extensively in applications that are high in terms of speed and load size. Liquid lubricants are the most dominant type of lubricant in the market. Liquid lubricants comprise base oil and some additives. The various types of liquid lubricants are as follows.

2.4.1 Mineral Oils

Mineral-based lubricants are extracted from crude oil. Mineral oil lubricants are of four types. The first type is paraffinic oil. It has good resistance to oxidation. It exhibits good thermal stability, is less volatile, and has a high flash point. The second type of mineral oil lubricant is naphthenic oil. This type of lubricant is good for low-temperature applications. It has a lower flash point than paraffinic oil lubricant. When naphthenic oil lubricant is burnt, soft deposits are formed which in turn lowers the abrasive wear. The third type of mineral oil lubricant is multi-grade oil. It is made by adding polymers in mineral oils, thus enhancing the viscosity index of the lubricant. These lubricants have different grade levels whereby a specific grade of lubricant oil can offer optimal performance in low or high temperatures.

Lastly, synthetic oil is another type of mineral oil lubricant. This type of lubricant was created to withstand harsh operating conditions. Jet engines use synthetic lubricants. These lubricants are expensive but they can withstand high levels of heat and stress. Some commonly used synthetic oils are esters, silicon, polyglycols, perfluoropolyalkylether, and perfluoropolyethers.

2.4.2 Vegetable Oils

Oil-based lubricant usually made from edible and non-edible oils is known as vegetable oil lubricant. Vegetable oil contains more natural boundary lubricant than mineral oil. However, vegetable oil lubricant is less stable than mineral oil lubricant at high-temperature ranges. To boost sustainability, different types of edible vegetable oils (soybean, rapeseed, sunflower, olive, coconut, castor, etc.) are used in various applications. Further, due to the immense gap in demand and supply of edible vegetable oils, non-edible oils-based lubricants such as karanja, linseed, cottonseed, Jatropha, etc. are used.

2.4.3 Animal Oils

Fats extracted from fish and animals are the sources of animal oil. Animal oil is sometimes also known as fixed oil. It is added to mineral oil in order to improve the film-forming ability of the mineral oil. Animal oil does not volatilize. The main drawback of animal oil is its availability.

2.4.4 Water-based Lubricants

Water has a natural cooling effect. The natural cooling effect reduces heat generation in the component, leading to considerably better efficiency. Furthermore, hydro lubricants minimize friction and wear, contributing to a considerably longer service life of the engine components.

Water is an additive and is, therefore, easier to meter, enabling easier cleaning during operation with a minimum of standstill and optimized working conditions since water, as a non-flammable application aid, reduces the fire hazard and is non-toxic when inhaled.

Water glycol is another non-flammable fluid commonly used in aircraft hydraulic systems. It generally has low lubrication ability as compared to mineral oils and is not suitable for high-temperature applications. It has water and glycol in the ratio of 1:1. Because of its aqueous nature and the presence of air, it is prone to oxidation and related problems. It needs to be added with oxidation inhibitors. Enough care is essential in using this fluid as it is toxic and corrosive toward certain metals such as zinc, magnesium, and aluminum. Again, it is not suitable for high-temperature operations as the water may evaporate. However, it is very good for low-temperature applications as it possesses high anti-freeze characteristics.

2.5 IONIC LIQUID LUBRICANTS

Many factors tend to influence the increased demand in recent years, including the state-of-the-art of effective and environmentally benign lubricants. More importantly, managing volatile demand specifically in the development of lubricant efficiently can be greatly significant to the rapid technological improvements in various engineering and manufacturing industries. To date, tailor-made ionic liquids investigated for application as lubricants have been known to play an important role in enhancing tribological interactions between sliding materials. Present interest concerns recent applications and emerging fields for the utilization of ionic liquids as new advanced lubricants. The rheological properties of ionic liquids, including their physical and chemical characteristics, have shown to be better than conventional lubricants.

Ionic liquids are organic salts and are defined as compounds that consist only of ions, where one or both of the ions are organic species. These ions are either positively or negatively charged and since at least one of them has a delocalized charge, they are poorly coordinated in their lattice structure which prevents them from forming crystal solids. Hence, with those attributed reasons, ionic liquids are liquid substances at temperature below 100 °C or even at room temperature. The cation/anion combination can exceed 1 million numbers of statistical prediction possibilities, which can be used to produce ionic liquids.

Currently, the main concern of industries is that the lubricants can be green which aims in pursuit of sustainability. If the sustainability model of green is considered, they can be more environmentally friendly, provide a better performance, and improve the economic bottom line. Thus, the industries want to proceed toward green in various aspects. The green include renewable, recyclable, reusable, non-toxic or less toxic, energy-conserving, and waste-reducing lubricants. However, most lubricants cannot be reused because of degradation and contamination, though some consumers have tried with limited success. For example, used lubricants are sometimes applied to moving chains. This is not considered the best lubrication practice, but success varies depending upon the condition of the used lubricant. Another reuse for lubricants is that they are collected and burned as heating fuel oil. The fuel is needed as an energy source, so this approach is greener than dumping into a landfill or pouring into the environment.

Industrial lubrication can be extremely green through responsibly planned purchasing, storage, use, and disposal, and challenging the limited regulatory view of green lubricants that fails to consider longer life of lubricant and component and decreased energy use. When properly formulated and used, most lubricants can last longer, therefore generating less waste and can be considered environmentally friendly. In the past, environmentally acceptable lubricants were made from bio-based materials or were biodegradable, most formulated using vegetable oil-based fluids. They typically became jelly-like at low temperatures and oxidized rapidly at operating temperatures. They were also more expensive. This meant that for a user to employ green lubricants, they had to pay more for a product that didn't perform as well. Today lubricants can be formulated using high-performance bio-based materials and meet the more traditional definitions of environmentally friendly, such as being biodegradable, low toxicity, and non-bio-accumulative. These biodegradable lubricants overcome the low- and high-temperature concerns of the past lubricants.

2.6 BIODEGRADABLE LUBRICANTS

In the current scenario, biodegradable lubricants become considerably important to the environment and are usually less toxic to natural habitats, protected species, land or water, and also less flammable. Biodegradable lubricants are either synthetic (base oil: synthetic esters and polyalkylene glycols) or derive from vegetable oils, and degrade significantly faster than mineral oils such as petroleum, reducing rapidly to components that are more readily broken down by natural micro-organisms. Biodegradable lubricants have the molecular ability to be degraded biologically (i.e. by the action of biological organisms). Bio-based lubricants may also be synthetic esters, which are partially derived from various natural sources such as solid fats, waste materials, and tallow. In general, mineral oils tend to be less biodegradable than esters. However, not all esters perform equally, with ester-type and branching causing significant variations in biodegradability. Most oils taken directly from animal and vegetable sources do not yield stable lubricants. It is the instability that makes them highly biodegradable, an environmental advantage. Due to the low toxicities of the above-mentioned base oils, aquatic toxicity exhibited by lubricants formulated from them is typically a consequence of the performance-enhancing additives or thickening agents used in the formulation, which can vary widely.

Biodegradable lubricants for different specifications such as hydraulic fluids are being developed with greater biodegradability and less toxicity that can operate in more environmentally sensitive areas. Almost one-half of the lubricants used across the globe are found to be lost due to accidental release or spills. This results in toxic oils that contaminate forests, fields, and waterways with limited detection. But, biodegradable lubricants while released into the environment can be broken down easily in soil and water.

In industry, biodegradable lubricants are evolving to the point where they can be used in almost every system that a conventional lubricant is traditionally found. Biodegradable lubricants can be categorized as sustainable because it is derived from renewable raw materials. These lubricants can help industries to meet customer sustainability expectations and environmental regulations.

2.6.1 Vegetable Oil-based Biodegradable Lubricants

Vegetable oils are being investigated as a potential source of environmentally favorable lubricants, due to a combination of biodegradability, renewability, and excellent lubrication performance. The majority of vegetable oils consisting of primarily triacylglycerides (also termed triglycerides) are glycerol molecules with three long-chain fatty acids attached at the hydroxyl groups via ester linkages. Biodegradable lubricants made from natural vegetable oils, such as rapeseed or sunflower seed, have come a long way in the last two to three years. In addition to their environmental benefits, vegetable oil-based biodegradable lubricants possess several advantageous performance qualities compared to mineral oil-based lubricants. They have a higher viscosity index, higher lubricity, and high flash point than conventional mineral oils. They perform well at extreme pressures and do not react with paints, seals, and varnishes. The limitations of vegetable oil-based biodegradable lubricants include poor performance at both low and high temperatures and oxidative instability (Habereder et al., 2008). Vegetable oils thicken more than mineral oils at low temperatures and are subject to increased oxidation at high temperatures, resulting in the need for more frequent oil changes. These shortcomings can be addressed with the use of selected additives for a formulation that is less susceptible to oxidative instability.

2.6.2 Synthetic Esters-based Biodegradable Lubricants

Synthetic esters-based biodegradable lubricants perform at a wide range of temperatures and exhibit high viscosity index, high lubricity, provide corrosion protection, and have high oxidative stability (Habereder et al., 2008). Because they contain biobased material, many synthetic esters satisfy testing requirements for biodegradability and aquatic toxicity, although they tend to be less readily biodegradable than pure vegetable oil-based lubricants. Synthetic ester-based lubricants can be more or less toxic than vegetable oil-based lubricants, depending on the aquatic toxicity of the additives used in the formulation. Some challenges do remain, and biodegradable lubricants still need to be monitored in the application as their service life can vary compared to mineral oils, depending on the type and application. However, more innovation in this space could mean synthetic ester-based lubricants can provide extended service life compared to mineral oils, which may go a long way to offset their initial cost.

2.6.3 Polyalkylene Glycols-based Biodegradable Lubricants

Polyalkylene glycols-based biodegradable lubricants are water-soluble and posses good lubricity, viscosity index, corrosion protection, and excellent low- and high-temperature viscosity performance among all of the types of bio-lubricants. For marine applications, these lubricants are attractive because, in addition to their high biodegradability, they retain their performance characteristics following water influx better than other biodegradable lubricants. The drawbacks of this type of lubricant include high cost, increased toxicity to aquatic organisms by directly entering the water column and sediments rather than remaining on the water column surface.

Table 2.1 summarizes the comparative environmental characteristics such as biodegradation rates, toxicity, and bioaccumulation potential of different lubricant base oils (Environmentally Acceptable Lubricants, 2011; Mudge, 2010). The biodegradability of a lubricant reflects that of the lubricant's base oil, the degree of toxicity, is usually an effect of the performance-enhancing additives (or thickening agents) within the formulation. The base oils that degrade quickly are considered

TABLE 2.1 Comparative environmental characteristics of different lubricant base oils

LUBRICANT BASE OIL	*SOURCE OF BASE OIL*	*BIODEGRADATION RATE*	*BIOACCUMULATION POTENTIAL*	*TOXICITY*
Mineral oil	Petroleum	Persistent/inherently	Yes	High
Vegetable oils	Naturally occurring vegetable oils	Readily	No	Low
Synthetic ester	Synthesized from biological sources	Readily	No	Low
Polyalkylene glycols	Petroleum-synthesized hydrocarbon	Readily	No	Low

more preferable than those that do not rapidly degrade, although there might be a trade-off with regard to the depletion of oxygen during compound metabolism. The compounds that do not bio-accumulate and are relatively less toxic are considered more preferable than those that bio-accumulate and have higher toxicities. Recently, a majority of mineral oils have the lowest biodegradation rate, a high potential for bioaccumulation, and measurable toxicity toward micro-organisms. On the contrary, the base oils derived from vegetable oils and synthetic esters degrade faster, have a smaller residual, do not bio-accumulate appreciably, and have lower toxicity to marine organisms. Polyalkylene glycols-based lubricants are also generally biodegradable and do not bio-accumulate, but some glycols may be more toxic due to their solubility in water. From Table 2.1 it is inferred that lower environmental impacts will take place if a greater proportion of base oils are produced from biologically-sourced materials.

At present, biodegradable lubricants are particularly used in the following applications:

- Total loss systems (chainsaw lubricants, corrosion preventatives, mould release oils).
- Hydraulics for excavators working in environmentally sensitive sites.
- Water pumps and grease applications where release into the environment is unavoidable.
- Smaller and efficient equipments operating at high speeds, temperatures, and pressures.

Innovation in biodegradable lubricants could convert into high-performance products suitable for use in environmentally susceptible fields.

2.7 NANO LUBRICANTS

2.7.1 Solid Nanoparticles-based Lubricants

Solid nanoparticles (which are the so-called zero-dimensional nanomaterials) blended in vegetable oils play a significant role to reduce friction and cutting temperature at contact surfaces. The reduction in temperature is mainly due to the large thermal conductivity of nano-fluids, which provide both cooling and lubrication action at the contact interfaces by rolling, self-mending/repairing, and tribo-film mechanisms. Researchers have used a wide variety of nanoparticles as solid lubricants and indicatively such particles include graphite, molybdenum bisulfide (MoS_2), titanium dibromide (TiB_2), calcium fluoride (CaF_2), cerium fluoride, boric acid (H_3BO_3) boron nitride, talc, tungsten disulfides (WS_2), aluminum oxide (Al_2O_3), silicon dioxide (SiO_2), and single- or multi-walled carbon nano-tubes (CNTs).

Further, two-dimensional (2D) nanomaterials are recently developed as additives to improve the friction reduction and anti-wear properties of base lubricants. 2D nanomaterials such as graphene, nanofilms, nanolayers, nanocoatings, etc. are widely used for tribological applications due to their unique molecular structure and lubricating properties. High strength between atoms within the same layer and low interlayer shear strength makes them excellent candidates for lubricating dual rubbing surfaces. Graphene nanomaterials as a solid lubricant have attracted the attention of the scientific community in tribological applications throughout the world. Owing to the sp^2 honeycomb structure along with very strong covalent bonds graphene nanomaterials show exemplary properties such as unique mechanical strength, exceptional lubricity, and superior heat transfer rate. Also, has the ability to separate the contact and to form a protective film on the surface. Other 2D nanomaterials such as phosphate, silicate, and oxide with layered structure have been newly explored as lubricant additives. The above nanomaterials could be useful to high-performance lubricating systems and to increase the energy efficiency and reliability of existing mechanical systems.

2.7.2 Water-based Nano Lubricants

In the current decade, water-based nano lubricants have emerged as promising eco-friendly lubricants compared to petroleum-based liquid lubricants. Water-based nano lubricants are formulated by dispersing nano additives into the water, which combines excellent cooling capacity of water with excellent lubricity contributed by the nano additives. The use of water-based nano lubricants not only provides protection against friction and wear between the tool and the workpiece during the machining process, but also improves overall quality of the product. This will show great potential in engineering applications. The tribological behavior and lubrication mechanism of nanomaterials as nano additives in water have recently been explored by researchers. The nano additives can improve the tribological performance of the lubricant in terms of friction, wear, oxidation resistance, corrosion resistance, anti-foaming, etc. The different types of nano additives dispersed in water include pure metals, metal and non-metal oxides, metal sulfides, carbon-based materials, composites, and some others such as metal salts, nitrides, and carbides (Morshed et al., 2021).

An entire new segment of the lubricants industry exists called re-refiners. In the infancy of re-refining, waste oil collectors took spent lubricant back to their facility, removed the water, filtered out the solids, and resold it for various lubrication uses. Modern re-refiners do the same, but unlike their predecessors, they introduce it into a refinery process just like crude oil. After processing, new high-quality base oils are produced that have been found to be of equal or better quality to virgin base oils. These can be used to produce new lubricants, restarting the closed-loop process.

Through proper base fluid and additive selection, it is possible to formulate lubricant products that operate for extended periods of time under proper maintenance without needing to be changed. This will result in less lubricant purchased, less used lubricant disposed, less maintenance labor, and ultimately less financial resources spent. Using a reliable lubricant will result in improved environmental, economic, and social sustainability. Further, the industry manufacturing additives are working more closely with the lubricant industry to design additives that are suitable for improving the performance of biodegradable lubricants.

The industrial researchers inferred that lubricant formulations with 30% re-refined base oil and 70% virgin base oil would be a good assumption for a more general carbon footprint of lubricants (Stephan, 2020). Moreover, the advancement in lubricants has the potential to reduce emissions and by improving and using effective engine lubricants there would be savings in fuel of the vehicle.

2.8 SUMMARY

In this chapter, the broad classification of lubricants was highlighted. The detailed description of each type of generally used lubricants and their sub-types were briefly discussed. The concept of ionic liquid lubricant and its importance was briefly detailed. Further, the industrial significance of newly explored lubricants like biodegradable and nano lubricants followed by the conceptual realization and applications of their various types was discussed in the chapter.

REFERENCES

Ahmed, W., Mustafa, A., Dassenoy, F., Sarno, M., & Senatore, A. (2021). A review on potentials and challenges of nanolubricants as promising lubricants for electric vehicles. *Lubrication Science*. 10.1002/ls.1568.

Environmentally Acceptable Lubricants (2011). United States Environmental Protection Agency Office of Wastewater Management Washington, DC 20460.
Georgescu, C., Solea, L. C., & Deleanu, L. (2019). Additivation of vegetal oils for improving tribological characteristics. IOP Conference Series: Materials Science and Engineering, 514 012012.
Gow, G. (2009). Chapter 14, lubricating grease. In: Mortimer, R., Fox, M., & Orszulik, S. (eds) Chemistry and Technology of Lubricants, 3rd Edition. Springer, Dordrecht, Heidelberg, London, New York. 547 pp.
Habereder, T., Moore, D., & Lang, M. (2008). Chapter 26, eco requirements for lubricant additives. In: Rudnick, L. R. (ed) Lubricant Additive Chemistry and Applications, 2nd Edition. CRC Press, Boca Raton, FL. 790 pp.
Li, J. J., Ge, X. Y., & Luo, J. B. (2018). Random occurrence of macroscale superlubricity of graphite enabled by tribo-transfer of multilayer graphene nanoflakes. *Carbon*, *138*, 154–160.
Li, Q. Y., Zhang, S., Qi, Y. Z., Yao, Q. Z., & Huang, Y. H. (2017). Friction of two-dimensional materials at the nanoscale: Behavior and mechanisms. *Chinese Journal of Solid Mechanics*, *38*(3), 189–214.
Liu, L., Zhou, M., Jin, L., Li, L., Mo, Y., Su, G., Li, X., Zhu, H., & Yu, T. (2019). Recent advances in friction and lubrication of graphene and other 2D materials: Mechanisms and applications. *Friction*, *7*, 199–216.
Morshed, A., Wu, H., & Jiang, Z. (2021). A comprehensive review of water-based nanolubricants. *Lubricants*, *9*(9), 89.
Mudge, S. M. (2010). Comparative environmental fate of marine lubricants. Unpublished manuscript. Exponent UK.
Spear, J. C., Ewers, B. W., & Batteas, J. D. (2015). 2D-nanomaterials for controlling friction and wear at interfaces. *Nano Today*, *10*(3), 301–314.
Stephan, B. (2020). Sustainability and lubricants: The next steps into a sustainable future, Verband Schmierstoff - Industrie e.V. It is accessed through http://www.lube-media.com (accessed on 20 Dec 2021)
Wu, H., Jia, F., Zhao, J., Huang, S., Wang, L., Jiao, S., Huang, H., & Jiang, Z. (2019). Effect of water-based nanolubricant containing nano-TiO_2 on friction and wear behaviour of chrome steel at ambient and elevated temperatures. *Wear*, *426-427*(1), 792–804.

Advanced Lubrication Techniques in Nanotechnology

3

INTRODUCTION

Now a days, meniaturization of components are increasing due to which more sophisticated and lighter products are available in market. These components are the combination of mechanical engineering and electrical engineering. But due to the smaller size of the components, the various forces act on the surface that degrades the life of components. These forces are vandarwall force, adhesive force etc. These forces can be reduced by the application of lubricant but due to the size constraint, the bulk lubrication can not be used. To over come this problem, researchers have suggested self-lubricating coating, bulk solid lubricant coating, surface modification by physical or chemical means and texture on surface. These methods can improve the life of the components and also improves the efficiency and reliability of micro/nano devices without affecting the environment. The detail description of each methods are discussed in this chapter.

3.1 COATING

The coating is used to protect the surface from wear by reducing friction between mating surfaces. It can be in the form of a thin or thick film. Further, it can be

DOI: 10.1201/9781003201199-3

divided into the soft coating or hard coating. The soft coating has given the lower friction but the hard coating has given higher wear resistance. Therefore, to utilize the advantage of both types of coating, researchers have developed the sandwich of soft and hard coatings. These sandwich coatings are developed into the range of nanometer to a few micrometers and these are widely used in compact micro/nano devices. The most widely used sustainable coating material in micro/nano devices is diamond-like carbon (DLC; Seshan & Schepis, 2018; Swaminathan, 2017).

3.1.1 Thin Film Coating

In thin film, the coating thickness is in the range of nanometer to a few micronmeters. It can be performed by various techniques, which are discussed in the following sub-sections.

3.1.1.1 Physical Vapor Deposition (PVD)

The basic principle of the PVD technique is melting and vaporization. The process is performed under a vacuum. It involves the basic four steps i.e. evaporation, transportation, reaction, and deposition. It improves the hardness, wear, and oxidation resistance of substrate due to which the life of the components is increased. It is used in almost every type of inorganic material and a few types of organic materials. The high capital cost, high vacuum, high skilled worker, require high cooling arrangement, and low rate of deposition are the limitations of this technique. It has a wide range of applications such as automobiles, aerospace, cutting tools, surgical tools, etc. Classification of PVD is discussed next.

3.1.1.1.1 Evaporative Deposition

In this technique, the targeted material is kept over the crucible and starts heating under high vapor pressure. Due to which the material melts and starts vaporizing. The vaporized atoms fall over the substrate surface kept on the top and start the condensation, causing the deposition on substrate surface.

3.1.1.1.2 Electron Beam PVD

In this technique, the high-energy electron beam is used to melt the targeted material under a high vacuum. The melted material is vaporized and deposited on the substrate surface.

3.1.1.1.3 Sputtering

This technique was developed by Langmuir in 1920 to deposit the thin film of metal such as Ni, Co, Au, Al, Ti, etc. on substrate. In this technique, the high energy of ions is bombarded on the targeted materials under high vacuum. After sufficient energetic particles on the surface, the atoms ejected and deposited on substrate. It is carried out using a top-down approach (means targeted material is kept on top and substrate is kept on bottom). It can produce a uniform thickness. This technique is widely used in microelectronics for decorative and protective coatings, pattern generation, and surface hardening.

3.1.1.2 Chemical Vapor Deposition (CVD)

In this technique, the reactant gases are introduced into the reaction chamber through forced convection that reacts with the heated substrate surface and forms a non-volatile solid film over the surface of substrate. The generated byproducts are further diffused into the main gas stream and transported through forced convection from the exhaust system. It consists of the system for gas delivery, reaction chamber, energy source, vacuum system, and process control equipment and exhaust system. This process can produce the deposition at a very high rate and also deposit materials that are very hard to evaporate. Further, it can grow the epitaxial thin film with good reproducibility. The corrosive and toxic gases, high temperature, and complex phenomena are the main limitations of this process. It is widely used to deposit the thin film in semiconductors and related devices, such as integrated circuits, sensors, and optoelectronic devices. It can also be used for the development of composite coating and powder production.

3.1.2 Thick Film

When the thickness of the film is greater than 200 micronmeter, it is known as a thick film. It is developed through spin coating. Spin coating is carried out in two stages. The parameters of spin coating are varied according to the application and thickness required. The purpose of the spin coater is to make coating thickness homogeneous by applying centrifugal force. The sample is held through the vacuum and released from a sample disk by releasing the vacuum from the vacuum releasing knob. If the thickness is more, it is known as bulk composites. Nowadays, researchers have tried to develop three-dimensional (3D) structure for various applications in nanotechnology to develop micro/nano devices.

3.2 TRIBOLOGY IN NANOTECHNOLOGY

The study of tribology is very important in the era of nanotechnology because the tribological behavior in microscopic scale is very much different from the macroscopic scale. In macroscopic scale, Amonton's law of friction which states coefficient of friction is independent of contact area and applied load holds good but in the case of microscopic/nanoscopic scale, this law of friction fails because in microscopic/nanoscopic scale surface forces become extremely dominant in comparison to the applied force. One of the important factors that influences the operations of a micro/nano device is the surface area to volume ratio of its components. The surface area to volume ratio is least for spherical objects but increases rapidly as the material is flattened into a thin wafer (Labianca et al., 1993). Hence, it is expected for micro/nano electromechanical systems (MEMS/NEMS), which are mostly made by changing thin flat wafer pieces into desired shapes, that the surface area to volume ratio is very high. Hence, surface forces become significant causing very high levels of adhesion and friction. High friction also leads to high wear because of plastic deformation and fracture at the interface where two surfaces meet. High adhesion can also cause stiction between components, thus causing failure of the MEMS/NEMS. Therefore, it is of utmost importance that researchers develop SU-8 polymer further to enhance its mechanical and tribological properties for its application as structural material for MEMS/NEMS and another micro/nano systems (Bangert & Eisenmenger, 1996). The SU-8 resin was invented at IBM in 1988. It has several following advantages over other polymer materials used for micro/nano systems fabrication:

1. Good solubility
2. High transparency
3. Glass and film formation
4. Low glass transition temperature (T_g)
5. Highly viscous solutions
6. Ultra-thick layers up to 500 mm by single coating
7. Highly uniform coating
8. Good biocompatibility
9. Good mechanical properties (elastic modulus and hardness).

It has two drawbacks: poor tribological properties (coefficient of friction ~ 0.7 against itself and many other materials) and mechanical properties that are lower than the traditional MEMS material, Si (elastic modulus ~190 GPa and hardness ~ 158 GPa). The improvement of these two properties would make SU-8 a very useful alternate structural material for MEMS/NEMS applications. This can be achieved through various ways as discussed next.

3.2.1 Bulk Composite with Solid Lubricant

In this method, researchers have tried to develop the bulk composite of SU-8 polymer with the addition of various solid fillers. The solid fillers generate the tribofilm at the interface, causing the reduction of the friction coefficient. The tribotest is performed either on pin-on-disc or ball-on-disc tribo setup. For example, Jigute et al. (2006a) have added silica nanoparticles in the matrix of SU-8 in various compositions and tested on linear friction apparatus using spherical balls of 6 mm diameter at a constant applied load and sliding speed. They observe that the friction coefficient widely depends on the interface materials, and optimum concentration of nanoparticles gives the lower wear and friction coefficient. Further, heat treatment of surface also shows lower wear due to enhanced cross-linking of SU-8 molecules. In another study, the same author has developed the SU-8/slica nanoparticles composite on the aluminum, quartz, and silicon wafers using a photolithography technique and performed the tribotest on the same setup. They observed the lower friction coefficient and fivefold reduction in wear when compared to pristine SU-8. Further, researchers added the various solid fillers such as graphite, graphene, multiwall carbon nanotube, hexagonal boron nitride, talc in SU-8 and conducted the test on ball-on-disc tribo setup. Each researcher reported that optimum concentration of nanoparticles enhanced the tribological performance of bulk composite of SU-8 (Katiyar et al., 2016a, 2016c; Rathaur 2018, 2019). In another study conducted by Katiyar et al. (2016b), it was found that the hybrid fillers such as graphite and talc also provide the best tribological and mechanical properties compared to pure SU-8. The solid lubricant graphite and talc provide lower frictional property and higher wear resistance property, respectively. Therefore, to utilize both properties they added these materials together and conducted the ball-on-disc tribotest for 4 lakh cycles. A lower friction coefficient of ~0.2 was observed for SU-8/graphite (15wt%)/talc (15wt%). Further, the addition of graphite and talc in bearing balls is explored by Rathaur et al. (2020). It was observed that hybrid bearing ball can be used for low load application because it has shown the lowest friction coefficient without addition of lubricant. This research can also reduce the bulk lubrication in bearing application.

3.2.2 Surface (Molecular) Modification

This is another technique based on the principle of changing only the top surface with attachment of molecules by chemical or physical means. There are very few works observed on the surface modification of SU-8 using these two techniques.

3.2.2.1 Physical Surface Methods

Physical surface modification is employed by Singh et al. (2011) in a two-step method. In the first step, the SU-8 surface was given an exposure to oxygen plasma followed by dip-coating of perfluropolyether (PFPE) molecules. This surface treatment reduced the initial and steady-state Coefficient of Friction (COF) by ~ 4–7 times and 2.5–3.5 times, respectively, and wear life increased by more than 1000 times. This method provided a physical means to attach PFPE lubricants molecules onto SU-8 surface. The OH functional groups of PFPE tend to form bonds with the functional groups created on the SU-8 surface during oxygen plasma treatment.

3.2.2.2 Chemical Surface Methods

Chemical surface modification has also been performed by Singh et al. (2011) in which the SU-8 surface is treated with ethanolamine-sodium phosphate buffer solution. It is followed by dip-coating with PFPE in a dilute solution. The chemical reaction scheme in this treatment is explained as follows. The amine group of ethanolamine and epoxy group of SU-8 reacts to form a chemical bond. In the next step, the polar hydroxyl (-OH) group on the other end of ethanolamine reacts with the -OH functional group of PFPE (Saravanan et al., 2013). This technique also gives a similar enhancement in the coefficient of friction and life of components as it is obtained from the physical surface modification.

This is because the top layer in both cases is made up of the same PFPE molecules. Physical and chemical bonding with SU-8 makes the surface wear durable. The physical surface modification has also been tried on PMMA surface with very improved results (Henry et al., 2000).

3.2.3 In-situ Lubrication

This technique is new now widely utilized to enhance the performance of micro/nano devices. In this technique, an appropriate concentration of well-known sustainable lubricant is mixed homogeneously in the matrix of SU-8. In most of the cases, PFPE is mixed in SU-8 matrix before further processing. For proper homogenization of PFPE lubricant droplets, an ultrasonic homogenizer is used. Then, it is cured under the exposure of ultraviolet rays. For example, Katiyar et al. (2016a) investigated the effect of lubrication properties on SU-8 by varying the lubricant base oil SN-150 and PFPE. They obtained that both mechanical and tribological properties improved simultaneously by 9–10 folds and 5–6 times, respectively in comparison to pure SU-8 composite. In addition, thermal property and wettability are also improved by the addition of liquid fillers in SU-8. In another study, Katiyar and Samad (2021) studied the various composite coatings over silicon wafer. They found that the micro- to nano-size droplets of PFPE lubricant are trapped in the matrix of SU-8 and it is properly cross-linked. These trapped droplets act as reservoirs of the lubricant at the interface. It is observed from this work that the initial wear occurs on the surface followed by the opening up of the reservoir due to which PFPE droplets came at interface. The presence of PFPE lubricant at the interface will reduce friction and the wear life is increased by several orders of magnitude. Tribological analysis was performed by using a nano tribometer at 200 mN normal load and 0.0314 m/s sliding speed for 500 cycles and the obtained frictional data are shown in Figure 3.1. The mechanism of in-situ lubrication was found to be of boundary and mixed lubrication where there is partial formation of fluid film between the solid surfaces. Figure 3.2 shows optical images of the counterface ball surface. The presence of liquid lubricant is seen on the counterface surface.

Moreover, ionic liquid (IL) is also used as a sustainable lubricant due to which they (Batooli et al., 2015) developed the composite of SU-8 with varying concentrations of IL (1-Methyl-3-octylimidazolium hexafluorophosphate). However, the IL provided the same reaction in coefficient of friction over pure SU-8. It is only confined to the first approximately 1000 cycles. The friction coefficient quickly raised to higher value in the range of 0.2 and above, but wear is reduced by half over 1800 cycles of sliding.

The present literature on the in-situ lubrication of SU-8 shows that it has excellent potential to further reduce friction and increase wear life. Hence,

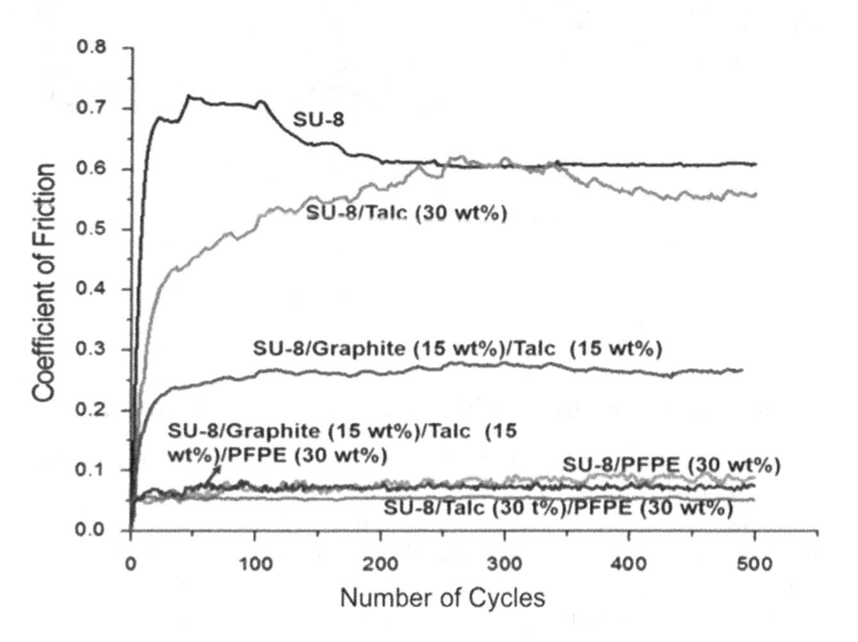

FIGURE 3.1 Plot of COF versus sliding cycles (Katiyar & Samad, 2021).

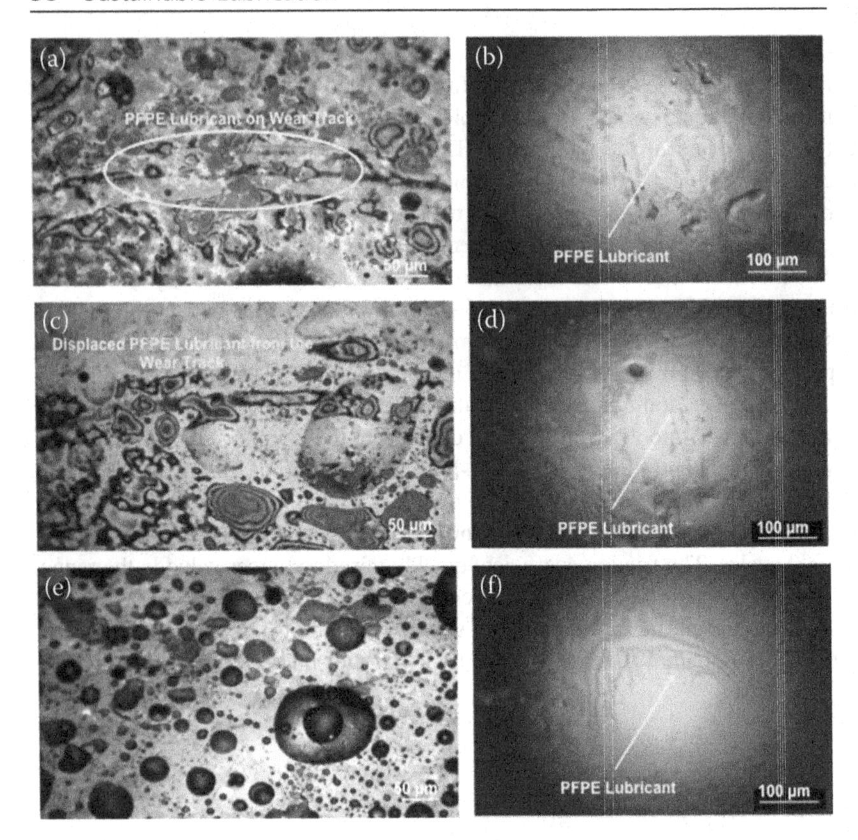

FIGURE 3.2 Surface and ball surface images after the wear tests for: (a, b) SU-8/PFPE (30 wt%); (c, d) SU-8/talc/PFPE (each 30 wt%); (e, f) SU-8/talc/Graphite/PFPE (each 15 wt% and 30 wt%) (Katiyar & Samad, 2021).

this method of improving tribological properties of SU-8 will be nowadays widely adopted by researchers.

3.2.3.1 Lubrication Mechanism in In-situ Technique

It can be observed that the PFPE molecules are uniformly dispersed on the surface as well as in the cross-section of SU-8 composites coating, due to which the surface exhibited a hydrophobic behavior. The possible lubrication mechanism at the interface is described through a schematic representation shown in Figure 3.3. Figure 3.3 (a) shows the tribo pair at the starting of the tribology test. As the test progresses, the top layer of PFPE molecules makes a tribo film, but it also gets evaporated due to frictional heating. As this

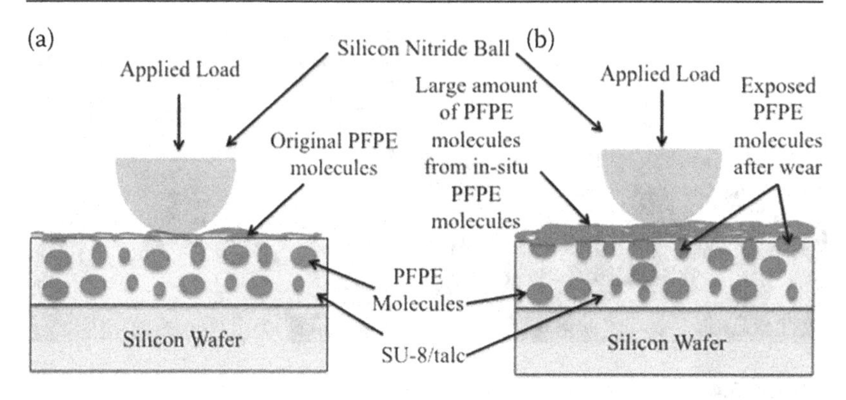

FIGURE 3.3 Possible lubrication mechanism for lower friction and wear (Katiyar et al., 2017).

happens, immediately, a new layer of PFPE molecules is exposed to the surface due to which a large amount of PFPE molecules are noticed at the interface as shown in Figure 3.3 (b). Hence, a continuous and stable tribo film of PFPE molecules is always present on the coating for the whole duration of the test due to which SU-8 composites with PFPE filler show lower friction and wear at interface. Hence, it is very helpful technique for reduction of friction in micro/nano devices (Katiyar et al., 2020).

3.2.4 Surface Texturing

Surface texturing is a very widely used approach before in-situ lubrication for reducing friction and wear. In this technique, the surface contact area is reduced, which results in lower adhesion and hence lower friction (Wakuda et al., 2003). One problem with textured surfaces is that the texture tends to get deformed or broken. If the texture material is a hard type, then the broken debris may even wear the remaining surfaces. Texturing of Si and polymer surfaces has been tried for reduction in friction. However, wear life was not increased appreciably. The nano-texturing effect provided by the oxygen plasma treatment of PMMA was found to increase the wear life of PMMA surface by many orders.

For example, Tay et al. (2011) fabricated rounded micro-dot pattern of SU-8 on silicon wafer of 2 mm × 2 mm dimension. The micro-dots of diameter 108.8 μm and height 1.14 μm were fabricated on Si wafers with different pitches and sliding experiments are conducted against flat Si wafers and Si_3N_4 balls. They observed that the micro-dot pattering can drastically reduce COF and there is an optimum pitch for which the friction coefficient is the lowest. This pitch length was found to be 150 μm for this case. It is also confirmed in this case that the wear

FIGURE 3.4 SEM images of SU-8, with and without diamond-like carbon (DLC) microstructures (a) connected, (b) isolated (Osborn et al., 2012).

life cannot be high despite low friction. The initial wear tends to accelerate the wear process in the presence of a texture. However, this effect is mitigated by providing a thin layer of PFPE lubricant on top of SU-8 micro-dots. Thus, texturing with top surface lubrication increased wear life by two to three orders of magnitude without failure.

Further, Osborn (2013) have fabricated, connected, and isolated micro structures (pillars) of SU-8, with and without DLC coating as the top layer. These structures are shown in Figure 3.4. They performed reciprocating tribological tests using 7-mm-diameter chrome steel ball as the counterface and obtained COF in the range of 0.8–0.4 and very less wear debris (1000 number of cycles).

Moreover, in the field of surface texturing, Myint et al. (2013) have fabricated 3D negative fingerprint and honeycomb textured surfaces on SU-8. They performed tribological tests against silicon nitride ball counterface on ball-on-disk tribometer. They have obtained the coefficient of friction of 0.08 for negative figure print as a compared to 0.2 for the untextured SU-8 surface in a rotational test at 100 mN normal load. The wear life of the negative figure print texture on SU-8 was highest compared with untextured and honeycomb texture when PFPE was also used as the lubricant layer.

3.3 LUBRICATION IN MICRO/NANO DEVICES

Lubrication of MEMS/NEMS has been a challenge because of the small size of MEMS/NEMS, nature of the structural material, Si, need for lubrication process to be compatible with the micro-fabrication processes, and no possibility of re-lubrication once the MEMS/NEMS is in service. Because of these constraints, MEMS/NEMS lubrication technology is still in an evolving period. A number of solutions have been proposed and tried in laboratories. Some of the important ones are gaseous lubrication by 1-pentanol hermetically sealed inside MEMS (Asay et al., 2008), use of self-assembled monolayers (OTS) (Satyanarayana & Sinha, 2005), localized lubrication of PFPE, etc. Though these methods are compatible with MEMS fabrication technique, their efficiency and cost-effectiveness vary. The new concept of in-situ lubrication, in comparison, is very simple and cost-effective. It does not require any extra step of lubrication during the micro/nano-fabrication. The method utilizes the changing of the bulk of the material which improves both the tribological as well as mechanical properties. As mentioned earlier, the in-situ lubrication works in the boundary or mixed lubrication regime. The mechanism of mixed lubrication is explained in the next section.

3.3.1 Mechanism of Mixed Lubrication

The mixed lubrication regimes tend to be effective in providing low friction, if there is formation and reformation of the boundary film. Formation and reformation of the boundary film is dependent on the role of absorption of additives. The additives are present in the lubricant as floating molecules. Hence, the process of formation and reformation of effective boundary layer is a diffusion-controlled phenomenon. Based on this concept, Albertson (1963) introduced an equation for the dynamic coefficient of friction, μ_v, at a given sliding velocity, v, as,

$$\mu_v = \mu_d - (\mu_d - \mu_\infty)\left[1 - e^{\left[\frac{-c}{v}\right]}\right] \quad (3.1)$$

where μ_d and μ_∞ are the coefficient of friction of fully damaged boundary film (substrate fully exposed) and fully effective boundary film (possible at low speed), respectively, and c is a diffusion parameter for the rate of diffusion of additives and molecules.

Based on the Eyring (1936) activated slip model. The potential barrier to slip can be overcome by the shear stress and thermal fluctuations. Hence, the net rate of passage of the molecules under the influence of external stress would be a sum of the forward and the backward movements of the molecules. The resultant rate of passage will be given as,

$$k_f - k_b = A\left[e^{-\frac{\left(E-\frac{\tau\varphi}{2}\right)}{kT}} - e^{-\frac{\left(E+\frac{\tau\varphi}{2}\right)}{kT}}\right] \tag{3.2}$$

where k_f and k_b are the rates of passage across the potential barriers, E is the energy barrier between atoms or molecules, τ is shear stress, φ is shear activated volume ($\lambda \times A$), A is the area over which the force act, λ is the distance moved by the molecules from one side to another across a potential barrier, k and T are the Boltzmann constant and absolute temperature, respectively.

The measure of slip is then given as the relative velocity v which is given as

$$v = \lambda\,(k_f - k_b)$$

$$v = A\lambda e^{-\frac{E}{kT}}\left[e^{\frac{\tau\varphi}{2kT}} - e^{-\frac{\tau\varphi}{2kT}}\right] \tag{3.3}$$

Since $\sin h\,(x) = \frac{(e^x - e^{-x})}{2}$

Equation (3.3) can be re-written as,

$$v = 2A\lambda\ e^{-\frac{E}{kT}}\sinh\left(\frac{\tau\varphi}{2kT}\right) \tag{3.4}$$

For low applied shear stress $kT > \gg \left(\frac{\tau\phi}{2}\right)$ and hence equation (3.4) is re-written as

$$v = A\lambda\,e^{-\frac{E}{kT}}\left(\frac{\tau\varphi}{2kT}\right) \tag{3.5}$$

Equation (3.5), provides a relation of the slipping or sliding speed within the lubricant as the applied shear stress is changed. When the applied shear stress is high

$$e^{\frac{\tau\varphi}{2kT}} \gg \gg e^{-\frac{\tau\varphi}{2kT}}$$

Hence, the equation (3.4) can be modified as:

$$v = A\lambda\,e^{-\frac{E}{kT}}e^{\frac{\tau\varphi}{2kT}} \tag{3.6}$$

Equation (3.7) shows that the speed increases exponentially with the applied shear stress. Briscoe et al. (1992) have shown that the potential barrier E actually changes with the hydrostatic pressure and the actual potential barrier is given as $E + p\Omega$, where p is the pressure and Ω is a pressure coefficient. Hence, equation (3.6) is then modified as below,

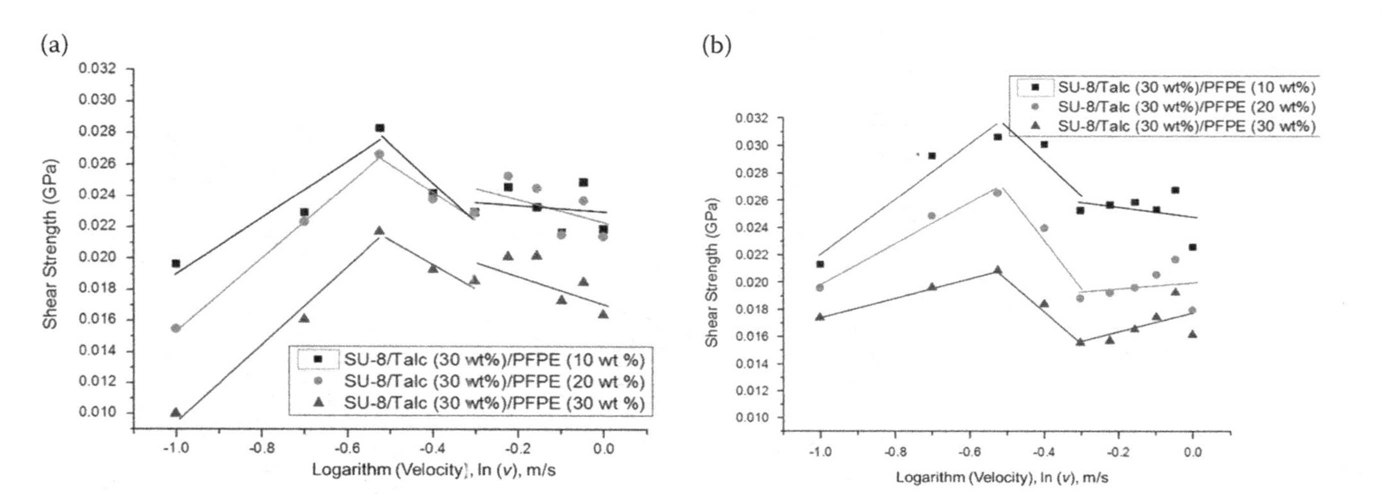

FIGURE 3.5 Shear strength at the interface with respect to logarithm (velocity) graphs for different SU-8 composites tested at (a) 2 N and (b) 4 N normal loads and varying sliding speed (0.1–1.0 m/s) (Katiyar et al., 2017).

$$v = v_0 e^{-(E+p\Omega)} e^{\frac{\tau\varphi}{2kT}} \quad (3.7)$$

where $A\lambda = v_0$, a velocity term. Equation (3.7) can be written as

$$\tau = \frac{kT}{\varphi} \ln \frac{v}{v_0} + \frac{1}{\varphi}(E + p\Omega) \quad (3.8)$$

Equation (3.8) can be given in the following form as,

$$\tau = \tau_0 + \theta \ln(v) \quad (3.9)$$

where $\tau_0 = \frac{1}{\phi}(E + p\Omega - kT \ln(v_0))$ and $\theta = \frac{kT}{\phi}$.

In equation (3.9), τ_0 and θ are constants for a given monolayer. Equation (3.9) states that there is a logarithmic relation between the shear stress and the sliding speed. Hence, a plot of τ versus ln(v) will give linear result. These results have been validated by Katiyar et al. (2017) in their work. They tested the SU-8/Talc/PFPE composites at varying speeds (0.1–1.0 m/s) and varying applied load (2 N and 4 N) on ball-on-disc tribotester. The obtained result of shear stress with logarithmic sliding speed is shown in Figure 3.5. The obtained results have shown that interfacial shear strength reduces with increase in the concentration of PFPE liquid filler. This is very significant as the success of in-situ lubrication lies in reducing the interfacial strength. The value of τ_o is very low for 30 wt% PFPE composite compared to those of 10 wt% and 20 wt% composites.

3.4 SUMMARY

In micro/nano devices the surface-to-volume ratio is very high due to which adhesion, Van der Waals forces, etc. play a very important role in performance of micro/nano devices. These forces wear the device very fast and reduce the life of components. It can be reduced either by surface coating or by bulk composite coating or by in-situ lubrication or surface texturing. Among all methods, in-situ lubrication, in comparison, is very simple and cost-effective. It does not require any extra step of lubrication during the micro/nano-fabrication. The method utilizes the changing of the bulk of the material, which improves both the tribological as well as mechanical properties. Further, the in-situ lubrication works in the boundary or mixed lubrication regime.

REFERENCES

Albertson, C. E. (1963). The mechanism of anti-squawk additive behavior in automatic transmission fluids. *ASLE Transactions*, *6*, 300–315.

Asay, D. B., Dugger, M. T., & Kim, S. H. (2008). In-situ vapor-phase lubrication of MEMS. *Tribology Letter*, *29*, 67–74.

Bangert, H., & Eisenmenger, C. (1996). Deposition and structural properties of two-component metal coatings for tribological applications. *Surface and Coatings Technology*, *80*, 162–170.

Batooli, L., Maldonado, S. G., Judelewicz, M., & Mischler, S. (2015). Novel SU-8/Ionic liquid composite for tribological coatings and MEMS. *Micromachines*, *6*, 611–621.

Briscoe, B. J., Thomas, P. S., & Williams, D. R. (1992). Microscopic origins of the interface friction of organic films: The potential of vibrational spectroscopy. *Wear*, *153*(1), 263–275.

Eyring, H. (1936). Viscosity, plasticity, and diffusion as examples of absolute reaction rates. *Journal of Chemical Physics*, *4*(4), 283–291.

Henry, A. C., Tutt, T. J., Galloway, M., Davidson, Y. Y., McWhorter, C. S., Soper, A. S., & McCarley, R. L. (2000). Surface modification of poly(methyl methacrylate) used in the fabrication of microanalytical devices. *Analytical Chemistry*, *72*(21), 5331–5337.

Jiguet, S., Judelewicz, M., Mischler, S., Hofmann, H., Bertsch, A., & Renaud, P. (2006a). SU-8 nanocomposite coatings with improved tribological performance for MEMS. *Surface and Coating Technology*, *201*, 2289–2295.

Jiguet, S., Bertsch, A., Judelewicz, M., Hofmann, H., & Renaud, P. (2006b). SU-8 nanocomposite photoresist with low stress properties for micro-fabrication applications. *Microelectronic Engineering*, *83*, 1966–1970.

Katiyar, J. K., Sinha, S., & Kumar, A. (2016a). Effects of carbon fillers on the tribological and mechanical properties of an epoxy based polymer (SU-8). *Tribology – Materials Surfaces & Interfaces*, *10*(1), 33–44.

Katiyar, J. K., Sinha, S., & Kumar, A. (2016b). Friction and wear durability study of SU-8 composites with talc and graphite as fillers. *Wear*, *362–363*, 199–208.

Katiyar, J. K., Sinha, S., & Kumar, A. (2016c). In-situ lubrication of SU-8/talc composite with base oil (SN150) and perfluoropolyether. *Tribology Letters*, *64*(1), 5.

Katiyar, J. K., Sinha, S., & Kumar, A. (2017). Lubrication mechanism of SU- 8/Talc/PFPE composite. *Tribology Letters*, *65*(3) 84.

Katiyar, J. K., Sinha, S., Kumar, A., & Hirayam, T. (2020). Tribological analysis of Tip-Cantilever made of SU-8/talc/PFPE composite. *Tribology – MaterialsSurfaces & Interfaces*, *14*(2), 92–101.

Katiyar, J. K., & Samad, A. Md. (2021). Physio- tribo-mechanical Properties of Polymer Composite Coating on Silicon Wafer, Tribology International. 165, 107307.

Labianca, N. C., Gelorme, J. D., & Lee, K. Y. (1993). High aspect ratio optical resist chemistry for MEMS application. Proceedings of 4th International Symposium on Magnetic Materials, Processes and Devices, Chicago, 386–396.

Myint, S. M., Minn, M., Yaping, R., Satyanarayana, N., Sinha, S. K., & Bhatia, C. S. (2013). Friction and wear durability studies on the 3D negative fingerprint and honeycomb textured SU-8 surfaces. *Tribology International*, *60*, 187–197.

Osborn, L. (2013). Enhanced tribological properties of surfaces patterned with SU-8/DLC microstructures. *Inquiry: The University of Arkansas Undergraduate Research Journal*, *15*(1), 69–86.

Rathaur, A., Katiyar, J., & Patel, V. (2019). Thermo-mechanical and tribological properties of SU-8/h-BN composite with SN150/perflouropolyether as filler. *Friction*, *8*(1), 151–163.

Rathaur, A., Katiyar, J., Patel, V., Bhaumik, S., & Sharma, A. (2018). A comparative study of tribological and mechanical properties of composite polymer coatings on bearing steel. *International Journal of Surface Science and Engineering*, *12*(5/6), 379–401.

Rathaur, A., Katiyar, J., & Patel, V. (2020). Tribo-mechanical properties of graphite/talc modified polymer composite bearing balls. *Material Research Express*, *7*(1), 015305.

Saravanan, P., Satyanarayana, N., Siong, P. C., Doung, H. M., & Sinha, S. K. (2013). Tribology of self-lubricating SU-8+PFPE composite based lub-tape. Procedia Engineering in Malaysian International Tribology Conference.

Satyanarayana, N., & Sinha, S. K. (2005). Tribology of PFPE overcoated self-assembled monolayers deposited on Si surface. *Journal of Physics D: Applied Physics*, *38*, 3512–3522.

Seshan, K., & Schepis, D. (2018). *Handbook of Thin Film Deposition*. Elsevier Inc.

Singh, R. A., Satyanarayana, N., & Sinha, S. K. (2011). Surface chemical modification for exceptional wear life of MEMS materials. *AIP Advances*, *1*, 042141.

Singh, R. A., Satyanarayana, N., Kustandi, T. S., & Sinha, S. K. (2011). Tribo-functionalizing Si and SU-8 materials by surface modification for application in MEMS/NEMS actuator-based devices. *Journal of Physics D: Applied Physics*, *44*, 015301.

Swaminathan, P. (2017). *Semiconductor Materials, Devices and Fabrication*. Wiley, India. 2017.

Tay, N. B., Minn, M., & Sinha, S. K. (2011). Tribological study of SU-8 micro-dot patterns printed on Si surface in a flat-on-flat reciprocating sliding test. *Tribology Letter*, *44*, 167–176.

Wakuda, M., Yamauchi, Y., Kanzaki, S., & Yasuda, Y. (2003). Effect of surface texturing on friction reduction between ceramic and steel materials under lubricated sliding contact. *Wear*, *254*(3–4), 356–363.

Sustainable Lubrication for Systems

4

INTRODUCTION

Most of the lubricants used in industrial systems are based on mineral oils and formulated with required additives for a variety of lubrication purposes. The purposes include minimal frictional heating, minimal wear, and enhanced component life by enhancing the service life. Presently, the global consumption of lubricants is approximately 37.4 million tons, in which automotive sector consume major amount of lubricants (almost 68%) and the rest 32% consumed by other industries. The 32% lubricants include 12% hydraulic oils, 15% metalworking and cutting fluid, 3% greases and 2% gear oil. The use of these lubricants and their indiscriminate disposal subsequently produce environmental problems like pollution in water and soil, and further, the degradation of raw materials in environment is of serious concern. Therefore, researchers are developed the alternate lubrication, which is biodegradble and sustaiable. The sustainable lubrication is also reduced the friction and wear between moving surfaces in contact through development of thin film layer without affecting the environment and human beings. The sustainable lubricant can also be solid, liquid, or plastic, with oil and grease. The basic function of sustainable lubricant is to reduce friction, prevent wear, protect equipment from corrosion, accomplish temperature and contamination, convey power, protact environment and provide a fluid seal to the meting components. Therefore, the sustainale lubrication is very important in all machining process link turning, facing, milling, etc. as well as in between two meting parts to reduce the bulk lubrication and protact the energy and

DOI: 10.1201/9781003201199-4

environment. Hence, the widely used sustainable lubrication system in industries are discussed in this chapter.

4.1 MINIMUM QUANTITY LUBRICATION (MQL)

Earlier, the metalworking lubricant is used in a bulk quantity that affected the environment and operator's health as well as skin. These problems can be reduced by the use of MQL techniques that have shown a remarkable reduction in the process cost, reduction in lubricant consumption, and environmental conscious cooling that leads to improved machining with reduced cutting forces and power consumption. MQL is a technique used for advanced machine process under controlled flow rate and controlled environment. It is also known as eco-friendly machining (sustainable machining) or micro lubrication or near-dry machining. In MQL machining, a lesser amount of properly selected lubricant is applied over the surface that results in impressive cooling and lubrication at the tool–chip interface (Boubekri & Shaikh, 2012). It also created an excellence in machinability characteristics with a drastic reduction of cutting tool temperature, chip thickening ratio, and wear of cutting edge. Apart from this in the conventional flood lubricating application, the formed chips are wet, and they must be dried and cleaned for further use, which increases the working time and cost. Due to which MQL appears to be the best and effective method compared to dry and flood machining (Bukane et al., 2020). Due to more advantages in MQL, the researchers are continuously working on improving the performance of this technique. The broad classification of MQL is shown in Figure 4.1.

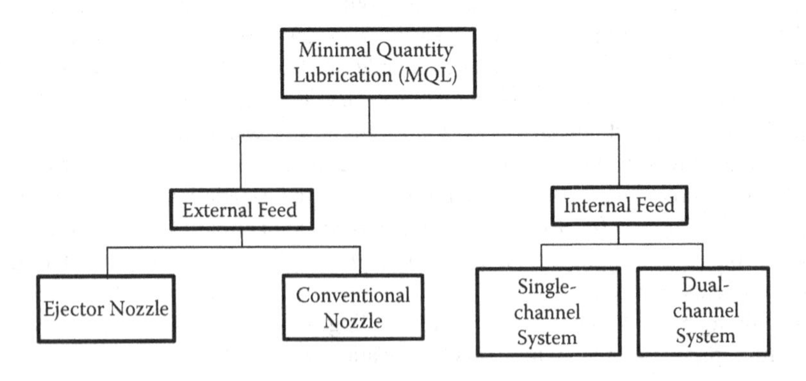

FIGURE 4.1 Types of minimum quantity lubrication.

4.1.1 External Feed

The external feed system is a spray type, which consists of a coolant tank or reservoir connected to tubes fitted with one or more nozzles. This system can be assembled near or mounted on the machine. The control valves are attached to adjust air and coolant flow for achieving balanced coolant delivery. It is inexpensive, portable, easy to maintain, and suited for almost all machining operations (Boubekri & Shaikh, 2012). In this system, the oil and air can be fed separately to the nozzle as in the ejector nozzle or mixed before feeding into the conventional nozzle (Abdelrazek et al., 2020). The schematic diagram of the external feed system is shown in Figure 4.2.

4.1.2 Internal Feed

The internal feed systems are available in two configurations as shown through a schematic representation in Figure 4.3. Based on the application, it

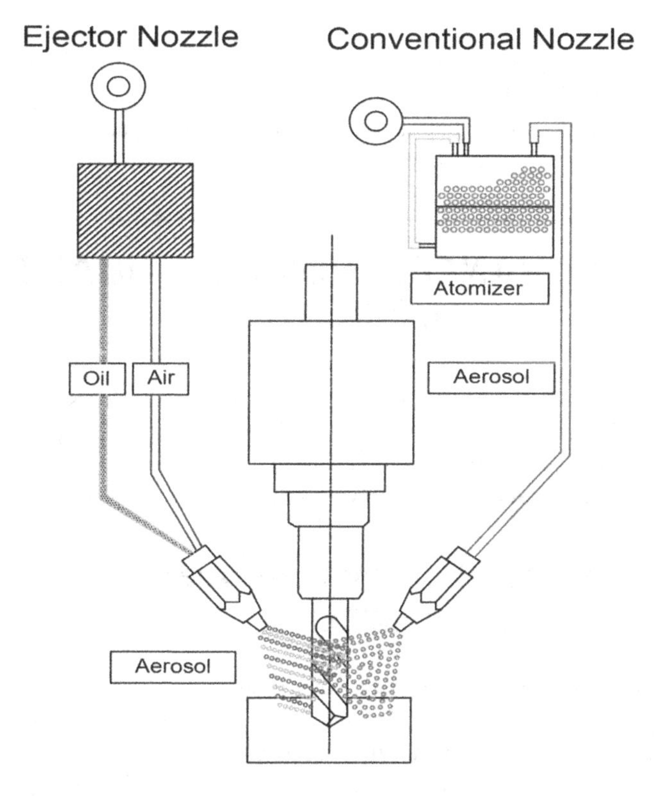

FIGURE 4.2 External feed system.

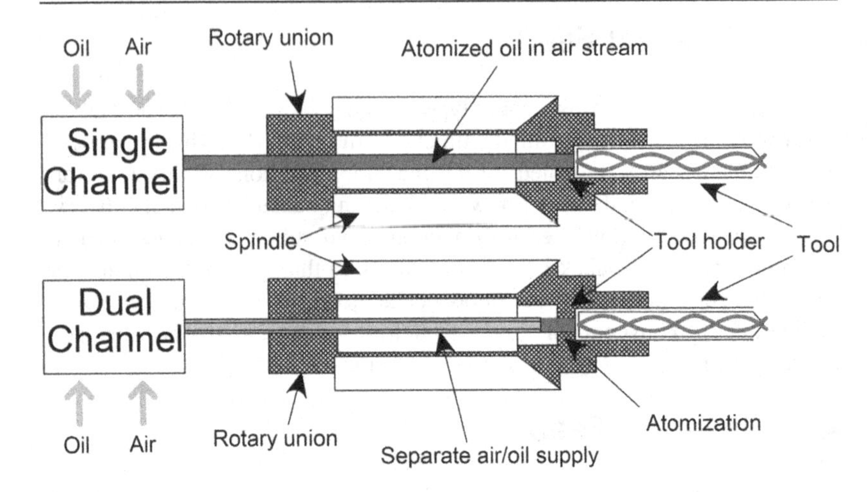

FIGURE 4.3 Internal feed/through tool.

created an air-oil mist. In a one-channel system, the oil and air are mixed externally, and flow through the spindle and tool toward the cutting zone. In the dual-channel system, the oil and air are supplied separately toward the tool holder where it is mixed before atomizing.

4.2 ADVANCED MINIMUM QUALITY LUBRICATION

To improve the further performance of machining, researchers have developed the advanced MQL system through which they obtained the higher tool life with lower cutting forces and enhanced surface roughness. These modifications are discussed in the following subsections.

4.2.1 Electrostatic Minimum Quantity Lubrication (EMQL)

EMQL is a distribution of lubricant using positive and negative electrostatic power supplies. Air and oil are stored in the reservoirs and then a needle electrode is connected to the nozzle with potential. Further, when a potential is applied, the electrostatically charged oil droplets are uniformly sprayed on the target of the machining zone. The performance of EMQL is altered by varying intensity and polarity of electrostatic voltage applied to the system. The study of EMQL revealed a decreased tool wear and surface roughness, with the reduction of 102% and 36%, respectively, when compared with the traditional MQL. The negative potential with low voltage and the positive potential with high voltage have shown better performance because of effective lubrication (Huang et al., 2018).

4.2.2 Nano-minimum Quantity Lubricant (NMQL)

The nano-based lubricants are dispersion of nanoparticles having size of less than 100 nm in any one dimension. The nanofluids can be metal, non-metals, or metal oxide dispersed in a base fluid (Bukane et al., 2020). The stable nanofluids enhance the thermal conductivity and lubrication. Nano-MQL can be used as both external feed and internal feed systems. Nanofluids are a smart fluid that contains nanoparticles in base oil which helps to provide superior lubrication and cooling properties as well as improve thermal conductivity and anti-wear properties (Gupta et al., 2019). Gupta et al. experimented the nanofluid for machining. The nanoparticles such as alumina, molybdenum disulfide (MoS_2), and graphite are used as additives in vegetable oil. Traditionally sonification technique is used to disperse the nanoparticles to the base fluid effectively. To enhance the dispersibility of nanoparticles the suitable surfactants are used. The cutting forces, tool wear, and surface roughness values were significantly affected with the addition of nanofluid lubricant for the machining process. In another paper nano MoS_2 is used as a lubricant along with conventional MQL method and the tribology properties of the lubricant were experimented. The nanoparticle in the colloidally dispersed fluid will fill in the microcavity and settle on the mating surface which reduces the friction and heat developed (Yücel et al., 2021). When compared to dry machining, the Ra values in the base fluid MQL approach improve by 8.41%, while the rate in the NFMQL strategy improves by 15.56%. The improved surface quality by MQL helps to form a tribo-film later at the tool chip interface, which leads to the protection of tool insert from heat and wear. Magnetic nanoparticles such as CuO are used to make magnetorheological fluid that will alter its rheology when a magnetic field was applied. In magnetorheological fluids, nanosized CuO is utilized as a binder to improve lubricity on the tool chip interference. These nanofluids are combined with Magnetorheological based Minimum Quality Lubrication (MRMQL) technique to achieve better lubrication. When compared to the MQL method, the cutting tool temperature was reduced by 9.36%, and the surface roughness was reduced by 26.58 % using MRMQL (Arul & Senthil Kumar, 2020).

4.2.3 Cold Air-assisted MQL

Cold air-assisted machining (CAAM), being environment friendly, is replacing fluid lubricant. The two types of cold air-assisted MQL for machining are vortex tube and cryogenic cooling. The coefficient of performance (COP) of a vortex tube and cryogenic cooling method is very important in refrigeration (Ginting et al., 2016). The vortex tube doesn't contain any moving parts and doesn't require electricity. In vortex tube producing cooled air, compressed air enters into the swirl chamber or vortex generator where the air revolves at higher revolutions per minute. It consists of two tubes, one is hot tube with a controlled valve and the other is cold tube. The lighter hot air in the outer vortex is vented out in the controlled valve, whereas the cold air is reversed and sent to the cold tube (Xue et al., 2013). It can produce cold air as low as –40 °C. A cold air-assisted minimum quality lubricant can produce higher pressure than a normal MQL method as well provide the necessary lubrication for maintaining tool life. The high-speed milling of Inconel 718 at a cutting speed of 210 m/min was studied in three different conditions such as cold air, flood lubricant, and dry cutting resulting in similar tool life (Kim et al., 2001). Due to the material asperities on the hills of the feed mark, an increase in surface roughness is noticed when using air-assisted drilling (Tasdelen et al., 2008). To overcome the above problem, the cryogenic machining method is used.

4.3 CRYOGENIC MACHINING

A cryogenic machining is a viable option due to its machining performance, non-toxicity, and environmentally friendliness. In cryogenic machining, some cryogenic gases such as O_2, H_2, CO_2, and N_2 are used. In conventional system, emulsion lubricants are used, whereas in cryogenic machining, a liquid or gaseous cryogenic fluid is used for lubrication. The wear characterization was studied for cryogenic cooling, Cryo-MQL against the conventional emulsion lubrication. The Cryo-MQL has shown better wear resistance and a lower coefficient of friction. Liquid coolant has shown better inhibition of heat than gas coolant (Verma et al., 2012). In addition, further hybridization of cryogenic machining techniques with MQL, EMQL, and nanofluids-based MQL (nMQL) can be beneficial for machinability improvements of difficult-to-machine materials and to extend the tool life.

4.3.1 Cryogenic-based Minimum Quantity Lubrication

Cryogenic treatment is the cooling process of materials at cryogenic temperature (below −1900 °C) to improve the hardness of materials, wear resistance, electrical conductivity, etc. Compared to other refrigerants, liquid nitrogen is more advantageous as it can be utilized at a wide temperature range. Application of cryogenic cooling has not only provided excellent lubrication but also helped in reducing the friction and heat generation at the tool-chip interface. The basic cryogenic system is shown in Figure 4.4.

The system can be able to supply continuous lubrication media into the liquid CO_2 flow where it can get the mixture of liquid CO_2 and lubricating media. The flow of the medium is precisely controlled by two separate need valves and flow meters (Grguraš et al., 2019). The liquid CO_2 mass flow rates were m_{LCO2} = 100 and 200 g/min. Oil volume flow rates were kept to V_{oil} = 20 and 60 ml/h, according to the usual MQL oil consumption. The droplets are found to be in the range of 2–10 micron. The small diameter of droplet exhibits better machinability as it is able to penetrate the cutting zone (Grguraš et al., 2019). In another work, a liquid nitrogen (LN_2) is used as the cryogenic coolant for MQL. The experiment is conducted with MQL, cryogenic machining, and Cryo-MQL. The results showed that Cryo-MQL improved surface roughness by 24.82% compared to cryogenic cooling, and tool wear was reduced 79.60% by the use of Cryo-MQL (Yildirim et al., 2020).

4.3.2 Ultrasonic Vibration-assisted Cryogenic-MQL (U-CMQL)

In this machining process, an ultrasonic vibration assists cryogenic MQL. The axial ultrasonic vibration produced by a piezoelectric crystal oscillator was applied to the cutting tool. Intermittent cutting enhances cutting fluid flow, which is supposed to offer lubrication and cooling at the cutting point. As a result of the reduced friction between the cutting tool and the workpiece, the thrust force is lowered, and the tool life is extended. Ultrasonic-vibration-assisted helical milling significantly lowered thrust force and the delamination size (Ishida et al., 2014). In another paper, the liquid N_2 (−196 °C) is used as

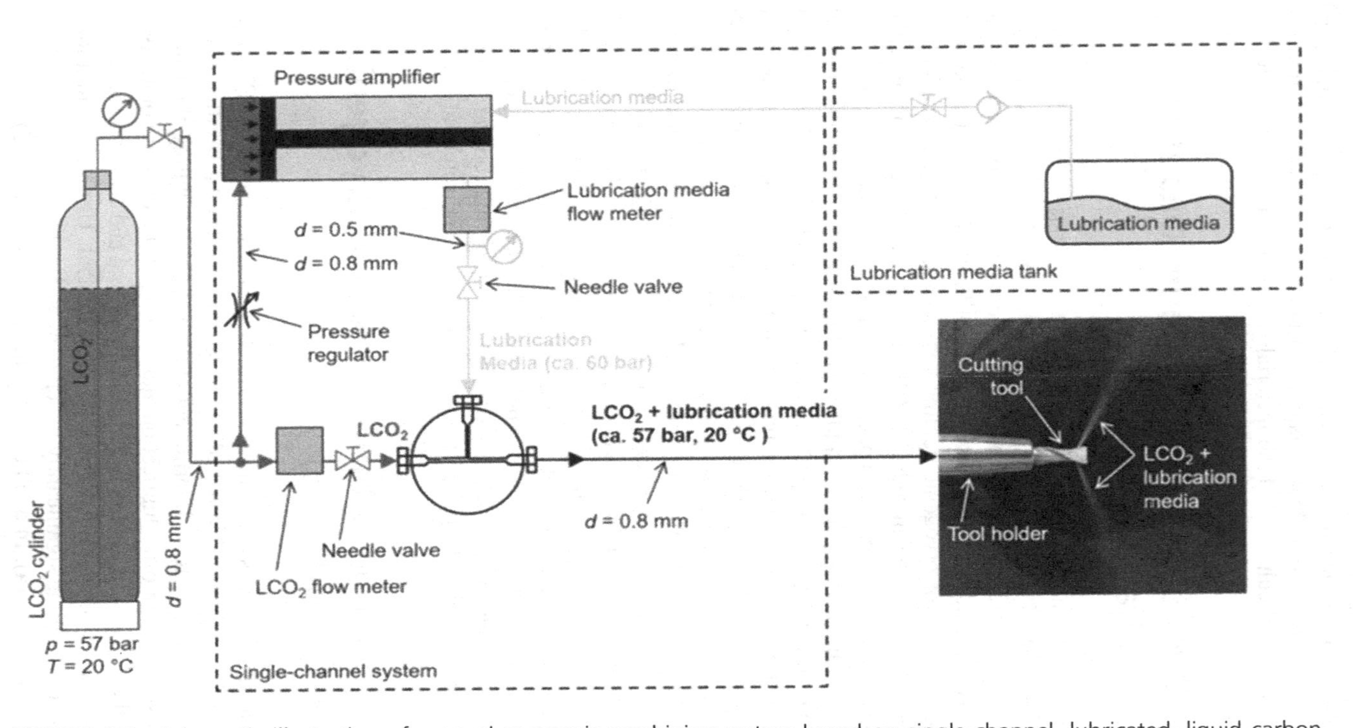

FIGURE 4.4 Schematic illustration of a novel cryogenic machining system based on single-channel, lubricated, liquid carbon dioxide (Grguraš et al., 2019).

the cryogenic cooling and was combined with ultrasonic vibration for the turning process.

4.4 GREASE LUBRICATION

The word grease is derived from the Latin word Crassus which means fat (Lugt, 2016). Grease is a solid to semi-fluid mixture of a non-flowable lubricant. Grease is the dispersion of solid lubricant in the base oil, a thickener, and additives. The rheology of the grease helps to stay on the surface for longer duration. The base oil for the grease lubrication is petroleum(mineral) oil, synthetic oil, or vegetable oil. Additives in grease lubrication play an important role to enhance the performance of the grease and helping to achieve desired properties. The main advantages of greases over oils lubricants are their ability to stay on in unsealed lubrication points, adapt over wider range of temperature and speed, better lubricity, higher corrosion protection properties, and a higher economic efficiency of application. The major drawbacks of greases are lower cooling capacity, a higher tendency to oxidation, and low possibility to use as centralized grease feeding system.

4.4.1 Ingredients in Grease Lubrication

4.4.1.1 Base Oil

The major composition of the grease lubricant is lubricant, base oil, rheology modifier, and additives (Ashraf & Aftab, 2012). The commonly used base oils are petroleum oils and synthetic oil. Petroleum oils are the byproduct during the distillation process of crude oil in various stages, each stage of oil exhibits different characteristic behaviors. The major component of crude oil is shown in Table 4.1.

4.4.1.2 Synthetic Oils

It is widely used as a dispersion medium for the production of grease. Synthetic oil can be able to operate over broad temperature and speed ranges, at high pressure, in a corrosive environment, etc. When petroleum base greases fail to meet the service requirements, the synthetic oils are used to produce specific greases. The major three synthetic oils used worldwide are polyalphaolefin (PAO) (45%), esters, including dibasic ester and polyol esters

TABLE 4.1 Major components of crude oil according to its specific temperature

NAME OF THE COMPONENT	*STATE OF MATTER*	*NUMBER OF CARBONS*	*BOILING RANGE (°C)*	*USES*
Residuals	Solid	Multiple-ringed compounds with 70 or more carbon atoms	Greater than 600	Coke, asphalt, tar, waxes; starting material for making other products
Fuel oil	Liquid	Long chain; 20 to 70 carbon atoms	370 to 600	Used for industrial fuel; starting material for making other products
Lubricating oil	Liquid	Long chain; 20 to 70 carbon atoms	300 to 370	Used for motor oil, grease, other lubricants
Diesel distillate	Liquid	Alkanes containing 12 or more carbons	250 to 350	Used for diesel fuel and heating oil, starting material for making other products
Kerosene	Liquid	Mix of alkanes (10 to 18 carbons) and aromatics	175 to 325	Fuel for jet engines and tractors; starting material for making other products
Gasoline	Liquid	Mix of alkanes and cycloalkanes (5 to 12 carbon atoms)	40 to 205	Motor fuel
Naphtha	Gas	Mix of 5 to 9 carbon atom alkanes	60 to 100	Intermediate that will be further processed to make gasoline
Petroleum	Gas	Small alkanes (1 to 4 carbon atoms); commonly known by the names methane, ethane, propane, butane	40	Used for heating, cooking, making plastics

(25%), polyalkyleneglycol (PAG) (10%) (Forbus, 2006). The other synthetic base oils are polyethylsiloxanes, polymethylsiloxanes, esters, synthetic hydrocarbon oils, polyalkylene glycols, polyphenyl ethers, fluoro-, and fluorochloro-carbon-based oils. Depending on the area and application suitable synthetic oil is used.

4.4.1.3 Polyalphaolefin

PAO is a type of molecularly designed base stock that has been optimized in terms of viscosity index, pour point, volatility, oxidative stability, and other critical lubricant base oil features. PAO is further classified based on the viscosities from 2 to 100 cSt at 100 °C. PAO usually has lower volatility, it is due to discrete carbon numbers with relatively long linear hydrocarbon branches. The carbon number lies between 30 and 42 (Forbus, 2006). PAO shows intrinsic oxidative stability and excellent response to antioxidant additive treatment.

4.4.1.4 Esters

Compared to PAO and mineral oil, ester fluids have a higher degree of polarity, superior additive solvency and sludge dispersancy, excellent lubricity, excellent biodegradability, and good thermal stability. The most used esters as synthetic base stocks are dibasic ester, polyol ester, and aromatic ester. Carboxylic diacids and alcohols are used to make dibasic esters. Dibasic ester is commonly used as a co-base stock with PAO to increase the solubility of lubricant and seal swell qualities. Polyols used to produce synthetic polyol ester base oils are pentaerythritol, trimethylolpropane, and neopentylglycol. Because of its branched structure, it shows good viscosity, better lubrication, and exhibits thermal resistance to cracking at 250 °C. Ester fluids are quite polar due to their high oxygen contents. They have high solubility for many commonly used additives. They also have a high solubility for polar acids and sludges produced during service by oxidation processes. Usually, 5–25% esters are used with PAO in finished lubricant formulations.

4.4.1.5 Polyalkylene Glycols (PAGs)

PAGs contain very high oxygen and have a hydroxyl end group(s). Water solubility and lubricity are enhanced by the presence of rich oxygen and hydroxyl end groups. As it is a water-based lubricant, it possesses fire-resistant properties. Starting ingredients for PAG production include ethylene oxide, propylene oxide, butylene oxides (BO), or mixtures of these epoxides. By selecting the right initiators, monomers, reaction conditions, and post treatments, PAG with a wide range of viscosities, pour points, water solubilities, and oil compatibilities can be prepared. BO-based PAGs have improved oil solubility and are not water-soluble. The PAG has a higher level of solvency and leaves a smaller footprint.

4.4.2 Soap Thickeners

During grease production, salts of higher fatty acids, soaps, and solid hydrocarbons; and less often, inorganic origin thickeners; and in small amounts, organic thickeners are used as a disperse phase of grease are called soap thickener. A soap thickener is important and plays a vital role in grease properties. Some well-known thickener in grease includes lithium, sodium, zinc, calcium, magnesium, strontium, potassium, barium, aluminum, lead, and other soaps. Signal or mixed thickeners are used in the production of grease, depending on the properties required. The soap thickener is further classified into anionic and cation soap thickeners. Lithium, sodium, potassium, calcium, barium, aluminum, and lead are widely used as soap cations in the dispersed phase of greases; hydroxides or other compounds of these metals are being employed to produce preformed soaps or as in-situ soaps formed during the grease production process. In the case of anionic soap thickener, it is extracted from vegetable oil, ground animal fats, and sea animal fats. Soaps, the dispersed phase of greases, are produced from a broad spectrum of fatty materials: fats of ground animals, vegetable oils, fats of fish and sea animals, commercial fatty acids, natural waxes, petroleum acids, and other commercial fatty materials.

4.4.2.1 Solid Hydrocarbons

Solid hydrocarbons are material that contains hydrogen and carbon in a linear, circular, or branched structure. Some solid hydrocarbons such as paraffin, petroleum jelly, and bitumen are found in petroleum, as well as natural waxes, etc., which are used as thickeners in the production of hydrocarbon greases serviceable up to 60–70 °C and offer good protective properties (Ishchuk, 2008).

4.4.2.2 Inorganic Thickeners

Solid inorganic materials have a high dispersity, hydrophobicity, ability to form the lubrication formulation, and act as a thickener in grease production. Inorganics thickeners such as modified silica (highly dispersed silica), clay minerals, lyophilic graphite, asbestos, and other minerals have the above properties. Clay minerals are widely used in grease production. Bentonite clays, palygorskite, kaolinite, and vermiculite are a few clay minerals used for grease production. These thickeners are also important both rheologically and economically. The other inorganic thickeners are oxides and hydroxides of various metals, such as hydroxides of aluminum, calcium, barium, strontium, and zinc, finely dispersed asbestos, mica, carbonates, sulfides, sulfates, phosphates of metals, etc.

4.4.2.3 Organic Thickeners

Organic compounds exhibiting the ability to form the grease texture, helping to form bonds between particles (structural elements) after their mechanical breakdown as well as showing good hydrophobicity, anti-corrosive properties are used for thickener to produce grease. For the practical application in grease production, the primarily used organics thickener is carbon black, polymers, and urea derivatives.

4.4.3 Additives and Fillers

Additives and fillers are materials that increase the performance characteristics of lubricants and reduce the price of lubricating materials. Additives can be readily soluble in the petroleum and other oils, dispersion media of grease. Fillers on other hand practically do not affect the structure formation. Additives and fillers help to maintain the colloidal stability of the product. Additives can significantly affect the process of structurization and the rheological characteristics of greases. The additives used in the production are based on the required specification such as oxidation inhibitors (antioxidants); anti-seizure; anti-wear and anti-friction; corrosion inhibitors and protective, viscous, thickening, and adhesive additives. Anti-foam additives are a polysiloxane, which improves the efficiency of processing and prevents spillage of materials from the reaction mixture. Fillers are usually in form of a powder that is highly dispersed oil-insoluble materials, which do not form a colloidal structure in greases but are an independent dispersed phase and improve the performance characteristics of greases. High lubricity, chemical, and thermal stabilities are the main causes of the extensive use of crystalline layered fillers featuring low coefficients of friction widely used as filler, such as molybdenum disulfide and diselenide, graphite, mica, talc, boron nitride, vermiculite, sulfides of some metals, and other materials. Solid high-polymer fillers such as polytetrafluoroethylene (PTFE), metal oxides, and metallic powders are employed as well. Molybdenum disulfide, graphite, mica, boron nitride, and polymers are used as filler materials.

4.4.4 Modern Grease Lubrication

Modern enhancements on the lubrication systems deal with both new developments in grease technology and in bearing design. The R0F+ and PDSC are two new breakthroughs methods for testing lubricating grease. Accelerated testing, which involves applying an extreme condition to shorten the testing time, should be related to the original parameter and be approached with caution. New testing methods need to be created to determine grease life at moderate temperatures (70 °C) in a reasonable amount of time. The contaminant particle and thickener particles traveling through the rolling element-ring cause the grease noise. A new test technology for grease noise has been developed. Even though more research works are going on in nanotechnology-related grease lubrication in past years still the grease market has not changed. Polymer grease and calcium sulfonate complex grease are two new advancements in grease thickening technology (Lugt, 2016). The nano-based grease lubricant such as carbon/ glass fiber-based polyamide 6,6, TiO_2/Carbon Nanotubes (CNTs), nanorod-Al_2O_3, graphene (Kunishima et al., 2021; Mohamed et al., 2020; Qiang et al., 2019), etc. The improvements in the rheological properties are observed after adding the nanorods. An addition of 0.3 wt% content of nanorods–Al_2O_3 to the grease composition exhibited the lowest average friction coefficient and wear scar diameter (Qiang et al., 2019). The addition of the nanoparticles to the system clearly shows better tribology properties and is not well commercialized.

4.5 SUMMARY

The sustainable lubrication system enhances the machining outputs, power consumption, machining cost, and environmental impact of the traditional machining processes. Many researchers observed that nano-fluid using the MQL technique and Cryo-MQL technique improves performance during machine and increases the tool life by reducing wear and improving lubrication properties. MQL is considered the most environmentally friendly machining process for turning, facing, and milling operations. More novel techniques are still under experimentation by most of the researchers to show little betterment in the lubrication system. In the early 1970s, the trend was to develop general-purpose grease. Modern enhancements on lubricating grease technology are designed for specific applications. The usage of synthetic-based grease lubrication protects the sliding surface from wear and extends the life of the sliding area. Greases are used in a place where it is very hard to re-lubricate, and sluggishness of lubrication is required. Further, the incorporation of the nanoparticles into the grease composition considerably increases the wear resistance, lubricity, and prolonged usage of lubricant. The biodegradable greases are made up of natural oil that degrades into water and soil. The food-grade lubricant is non-toxic lubricant that is used for food processing plants. The food-grade greases are made up of edible oils and antibacterial additives, which can prevent bacterial growth.

REFERENCES

Abdelrazek, A. H., Choudhury, I. A., Nukman, Y., & Kazi, S. N. (2020). Metal cutting lubricants and cutting tools: A review on the performance improvement and sustainability assessment. *International Journal of Advanced Manufacturing Technology*, *106*(9–10), 4221–4245. 10.1007/s00170-019-04890-w

Arul, K., & Senthil Kumar, V. S. (2020). Magnetorheological based minimum quantity lubrication (MR-MQL) with additive n-CuO. *Materials and Manufacturing Processes*, *35*(4), 405–414. 10.1080/10426914.2020.1732410

Ashraf, A. Al., & Aftab, A. Al. (2012). Distillation process of crude oil. *Qatar University*, *Bachelor Thesis*.

Boubekri, N., & Shaikh, V. (2012). Machining using minimum quantity lubrication: A technology for sustainability. *International Journal of Applied Science and Technology*, *2*(1), 111–115

Bukane, S., Shaikh, V. A., & Veerabhadrarao, M. (2020). Identifying optimization methods using MQL and Cryo-treatments for turning Inconel alloy with nanofluids. *Journal of Physics: Conference Series*, *1706*(1). 10.1088/1742-6596/1706/1/012219

Forbus, R. (2006). Practical advances in petroleum processing. *Practical Advances in Petroleum Processing*, *October 2007*. 10.1007/978-0-387-25789-1

Ginting, Y. R., Boswell, B., Biswas, W. K., & Islam, M. N. (2016). Environmental generation of cold air for machining. *Procedia CIRP*, *40*, 648–652. 10.1016/j.procir.2016.01.149

Grguraš, D., Sterle, L., Krajnik, P., & Pušavec, F. (2019). A novel cryogenic machining concept based on a lubricated liquid carbon dioxide. *International Journal of Machine Tools and Manufacture*, *145*(September), 1–6. 10.1016/j.ijmachtools.2019.103456

Gupta, M. K., Jamil, M., Wang, X., Song, Q., Liu, Z., Mia, M., Hegab, H., Khan, A. M., Collado, A. G., Pruncu, C. I., & Imran, G. M. S. (2019). Performance evaluation of vegetable oil-based nano-cutting fluids in environmentally friendly machining of inconel-800 alloy. *Materials*, *12*(7). 10.3390/ma12172792

Huang, S., Lv, T., Wang, M., & Xu, X. (2018). Enhanced machining performance and lubrication mechanism of electrostatic minimum quantity lubrication-EMQL milling process. *International Journal of Advanced Manufacturing Technology*, *94*(1–4), 655–666. 10.1007/s00170-017-0935-4

Ishchuk, Y. L. (2008). Lubricating Grease Manufacturing Technology, New Age International, ISBN 9788122416688.

Ishida, T., Noma, K., Kakinuma, Y., Aoyama, T., Hamada, S., Ogawa, H., & Higaino, T. (2014). New production technologies in aerospace industry - 5th machining innovations conference (MIC 2014) helical milling of carbon fiber reinforced plastics using ultrasonic vibration and liquid nitrogen. *Procedia CIRP*, *24*(C), 13–18. 10.1016/j.procir.2014.07.139

Kim, S. W., Lee, D. W., Kang, M. C., & Kim, J. S. (2001). Evaluation of machinability by cutting environments in high-speed milling of difficult-to-cut materials. *Journal of Materials Processing Technology*, *111*(1–3), 256–260. 10.1016/S0924-0136(01)00529-5

Kunishima, T., Nagai, Y., Bouvard, G., Abry, J. C., Fridrici, V., & Kapsa, P. (2021). Comparison of the tribological properties of carbon/glass fiber reinforced PA66-based composites in contact with steel, with and without grease lubrication. *Wear*, *477*(April). 10.1016/j.wear.2021.203899

Lugt, P. M. (2016). Modern advancements in lubricating grease technology. *Tribology International*, *97*, 467–477. 10.1016/j.triboint.2016.01.045

Mohamed, A., Ali, S., Osman, T. A., & Kamel, B. M. (2020). Development and manufacturing an automated lubrication machine test for nano grease. *Journal of Materials Research and Technology*, *9*(2), 2054–2062. 10.1016/j.jmrt.2019.12.038

Qiang, H., Wang, T., Qu, H., Zhang, Y., Li, A., & Kong, L. (2019). Tribological and rheological properties of nanorods–Al_2O_3 as additives in grease. *Proceedings of the Institution of Mechanical Engineers, Part J: Journal of Engineering Tribology*, *233*(4), 605–614. 10.1177/1350650118787403

Tasdelen, B., Wikblom, T., & Ekered, S. (2008). Studies on minimum quantity lubrication (MQL) and air cooling at drilling. *Journal of Materials Processing Technology*, *200*(1–3), 339–346. 10.1016/j.jmatprotec.2007.09.064

Verma, S., Kumar, V., & Gupta, K. D. (2012). Performance analysis of flexible multirecess hydrostatic journal bearing operating with micropolar lubricant. *Lubrication Science*, *24*(6), 273–292. 10.1002/ls

Xue, Y., Arjomandi, M., & Kelso, R. (2013). The working principle of a vortex tube. *International Journal of Refrigeration*, *36*(6), 1730–1740. 10.1016/j.ijrefrig.2013.04.016

Yildirim, Ç. V., Kivak, T., Sarikaya, M., & Şirin, Ş. (2020). Evaluation of tool wear, surface roughness/topography and chip morphology when machining of Ni-based alloy 625 under MQL, cryogenic cooling and CryoMQL. *Journal of Materials Research and Technology*, *9*(2), 2079–2092. 10.1016/j.jmrt.2019.12.069

Yücel, A., Yıldırım, Ç. V., Sarıkaya, M., Şirin, Ş., Kıvak, T., Gupta, M. K., & Tomaz, Í. V. (2021). Influence of MoS2 based nanofluid-MQL on tribological and machining characteristics in turning of AA 2024 T3 aluminum alloy. *Journal of Materials Research and Technology*. 10.1016/j.jmrt.2021.09.007, 15, 1688–1704

Application of Lubricants

5

INTRODUCTION

Presently, the global consumption of lubricants is approximately 37.4 million tons, in which automotive sector consume major amount of lubricants (almost 68%) and the rest 32% consumed by other industries. The 32% lubricants include 12% hydraulic oils, 15% metalworking and cutting fluid, 3% greases and 2% gear oil. In automobile, Engine is the most important component, and it requires continuous lubrication for smooth functioning. Engines consist of cylinder, piston, piston ring, connecting rod, gudgeon pin, crank, crank shaft, crank pin, inlet valve, exhaust valve, cam, and follower. During the movement engine part, it undergoes higher friction, resulting in less efficiency. The role of lubrication is not only to lubricate the engine parts, but it also helps to transfer the heat from engine cylinder to the oil sump. The lubricant forms a sacrificial thin film layer in between the sliding surface of engine that prevents them from contact and reduces the wear. Further, gear and bearings is other automobile parts, which undergo high pressure and wear.

5.1 GEAR

Gear is a circular component with teeth on the periphery of the wheel, which helps to transmit the motion from one place to another. Gears are broadly classified into spur gear, helical gear, herringbone gear, bevel gear, spiral gear, and hyperboloids and it is further classified into internal, external, and rack and pinion. Gear is manufactured either from metallic or non-metallic materials. Cast iron is widely used as the gear material due to its wear

DOI: 10.1201/9781003201199-5

resistance and self-lubrication properties. The other materials such as steel, alloys, bronze, etc., can be used to achieve the desired properties.

5.1.1 Nomenclature of Gear

Gear nomenclature is required to understand the working of gear and to design the gear for the desired purpose. In the transmission, usually, more than one gear will take place, the small wheel is a pinion gear. The *gear ratio* is the ratio of number of teeth in gear to pinion, as shown in Figure 5.1. The *pitch-circle* diameters of a pair of gears are the diameters of cylinders co-axial with the gears which will roll together without slip. The base circle is the circle from which the involute is generated. The *root diameter* is the diameter at the base of the tooth. The *center distance* is the sum of the pitch-circle radii of the two gears in mesh. The *addendum* is the radial depth of the tooth from the pitch circle to the tooth tip. The *dedendum* is the radial depth of the tooth from the pitch circle to the root of the tooth. The *clearance* is the algebraic difference between the addendum and the dedendum. The *whole depth* of the tooth is the sum of the addendum and the dedendum. The *circular pitch* is the distance from a point on one tooth to the corresponding point on the next tooth, measured around the pitch-circle circumference. The *tooth width* is the length of arc from one side of the tooth to the other, measured around the pitch-circle circumference. The *module* is the pitch-circle diameter divided by the number of teeth. The *diametral pitch* is the reciprocal of the module, i.e.

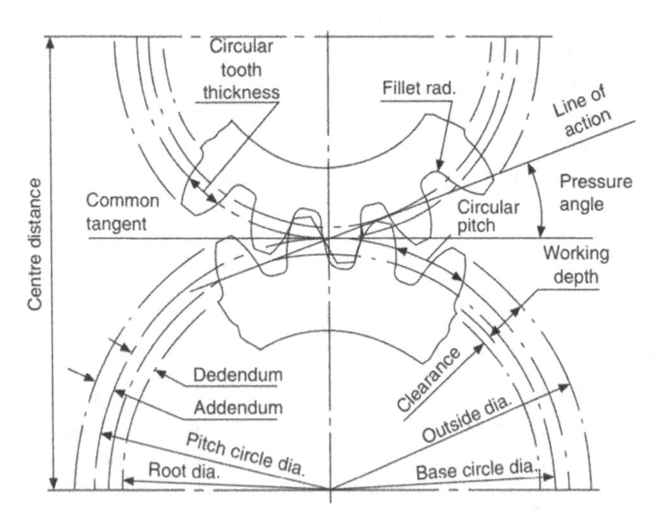

FIGURE 5.1 Nomenclature of gear (Simmons et al., 2020).

the number of teeth divided by the pitch circle diameter. The *line of action* is the common tangent to the base circles, and the *path of contact* is that part of the line of action where contact takes place between the teeth. The *pressure angle* is the angle formed between the common tangent and the line of action. The *fillet* is the rounded portion at the bottom of the tooth space.

5.1.2 Types of Gear

The gears are classified into different categories as discussed next.

5.1.2.1 Spur Gear

The spur gear is one of the most basic and widely used for the transmission of motion. It is used in cases where the shafts are parallel to each other. The spur gear's teeth are parallel to the rotational axis as shown in Figure 5.2. There is line contact between the teeth of two spur gears when they mesh. The spur gears are comparatively easy to design and provide constant velocity ratio. However, these gears are not suitable for perpendicular power transmission.

5.1.2.2 Helical Gear

Helical gear is quite similar to the spur gear; however, the teeth are usually inclined to 20 to 35 degrees as shown in Figure 5.2 (b). The helical gear is used when the shaft axis is parallel to each other. Helical gear operates smoothly due to its complete teeth engagement and produces less noise compared to spur gear. The sliding contact of helical gear produces high heat that reduces the efficiency and produces the high friction than spur gear.

5.1.2.3 Double Helical or Herringbone Gear

A double helical gear or herringbone gear has two helical gears facing opposite sides as shown in Figure 5.2 (c). Double helical gears help to reduce

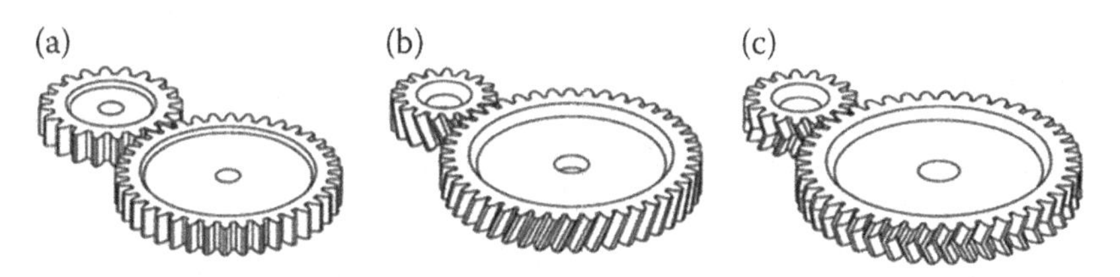

FIGURE 5.2 (a) Spur gear, (b) helical gear, (c) double helical gear or Herringbone gear (Watson, 1970).

the axial thrust caused by helical gears. The axial thrust caused by the right-hand teeth is canceled out by the left-hand teeth. Double helical gear is used for the transmission of high torque in heavy machinery, gas turbines, heavy-duty vehicles, torque gearboxes, etc. It is produced by high friction thus produces more heat as compared to other gears.

5.1.2.4 Bevel Gear

A bevel gear is formed on a conical surface, and they are used for transmitting motion between shafts with axes 90 degrees to each other. Also, the shaft axes can be said to be intersecting and non-parallel. A bevel gear is further classified into straight bevel gear and spiral bevel gear as shown in Figure 5.3 (a) and (b), respectively. Bevel gears are used in the differential gearbox of the automobile, hand drills, agricultural machines, printing machines, mining, robotics, cement mills, etc.

5.1.2.5 Rack and Pinion

The rack and pinion are a type of gear arrangement where the rotary motion of a pinion is converted into linear movement of the rack. The pinion is a small spur gear, and the rack is made up of gear teeth cut in a straight row on a flat surface as shown in Figure 5.3 (c). The applications of rack and pinion include steering mechanism of cars, lock gate control for canals, stairlifts, actuators to control the valves of pipelines, etc.

5.1.2.6 Worm Gear

A worm gear is a thread cut into a round bar, and a worm wheel is a gear that meshes with the worm at a shaft angle of 90 degrees as shown in Figure 5.4 (a). A worm gear can transmit high torque and produce self-locking. It occupies less space for the same speed reduction ratio as compared to other gears.

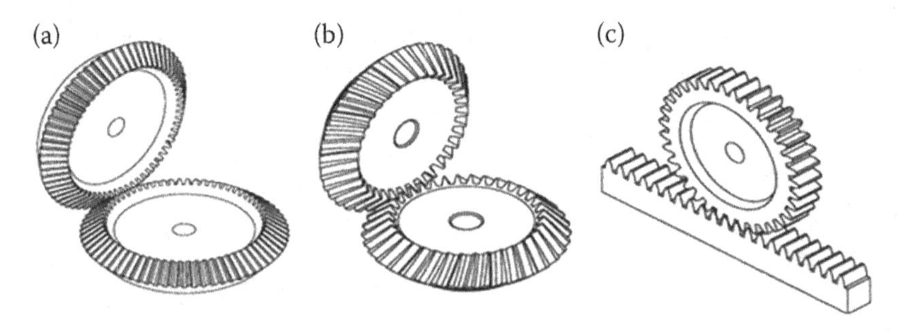

FIGURE 5.3 (a) Straight bevel gear, (b) spiral bevel gear, and (c) rack and pinion (Watson, 1970).

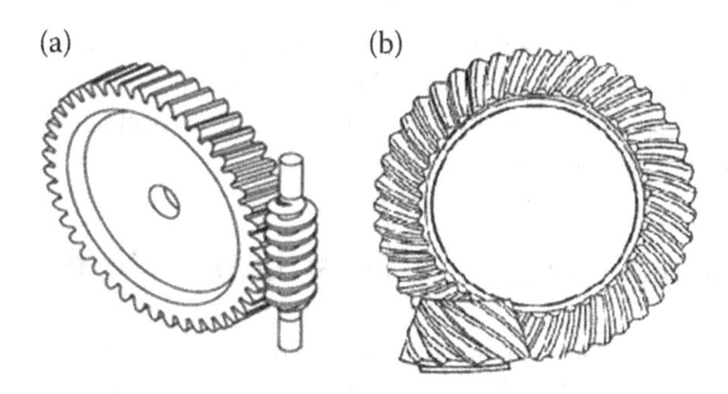

FIGURE 5.4 (a) Worm gear, and (b) hypoid gear (Watson, 1970).

5.1.2.7 Hypoid Gear

The shaft axes of a hypoid gear do not meet, unlike those of a spiral bevel gear. The pinion's axis is not parallel to the gear center as shown in Figure 5.4 (b). The shaft axes in hypoid gears are non-parallel and non-intersecting. In comparison to a spiral bevel gear, which is normally conical, this gear is in the shape of a hyperboloid. Hypoid gears are provided a higher speed reduction ratio that enables quiet and smooth operation.

5.1.3 Gear Lubrication

In comparison to belts and chains, gear is one of the most extensively used transmission elements because of no-slip between gears. Lubrication between the gear and its meshing pair has a big impact on operating life, durability, and efficiency (Ouyang et al., 2019). As per lubrication, the selection of material and its heat treatment process has played an important role to increase power density. For preparing the gear, the two most used heat treatment techniques are case hardening and nitriding. In the case of additive manufactured gears, a recent study highlighted the performance of gears produced by selective laser melting which retrieved mechanical strength that is generally lower than that of wrought material gears, especially under fatigue, owing to the induced porosities, and residual stresses (Croccolo et al., 2020). Increasing the efficacy and durability of gear lubrication reduces wear and friction acting between the teeth surface. Lubrication also helps to transfer the heat produced by friction, prevent power loss, and protect the parts. Based on the running speed the friction regime is classified into three levels: boundary friction, mixed friction, and fluid friction. In boundary friction, the tooth flanks are separated by a thin boundary layer of chemicals. In mixed friction, the tooth flanks are partially separated by a lubricating film, in this case, both dry and liquid friction can occur at the same time. In fluid friction, the tooth flanks are completely separated by the lubricating film that acts as elasto-hydrodynamic lubrication. At higher running speed, when a thickness of oil is greater than the roughness of the teeth then the elasto-hydrodynamic lubrication occurs.

5.1.4 Latest Trends in Gear Lubrication

The gear element requires continuous lubrication for extended usage and increased tool life. The lubricating oil on the gear surface is insufficient due to the increased operating power for heavy machinery resulting in lubricant starvation. This phenomenon can be avoided by using the surface coating and texturing on the gear parts. The surface texturing increases the surface area that facilitates heat dissipation and reduces the contact area, resulting in less friction. Zhang et al. (2019) proposed two methods of fabricating TiC coating one is direct cladding and the other is in-situ synthetic of TiCx coatings under low laser energy pulsed on the 40Cr gear steel surface. The study revealed the direct laser cladding of 50 wt% TiC coating has shown better wear resistance than the low-energy pulsed. Wang et al. (2020) demonstrated a nano-TiC functional gradient coating on the 40Cr gear steel surface using a laser cladding. The three layers cladding coating exhibited 50% reduction in coefficient of friction, a 40% reduction in grinding loss, and an increase in wear resistance. The microhardness is increased from 612 HV to 1088 HV. Liu et al. (2016) conducted a numerical method to analyze the coating on gear and found that the coating thickness and modulus ratio increase as the maximum film temperature increased. Increase in modulus ratio did not always affect the coefficient of friction under thermal conditions. Gupta et al. (2018) experimented the tribology behaviors of textured gearsets against conventional gearsets. The micro-cylindrical dimples of non-uniform diameter texture were made using a chemical etching process on the teeth of gear. Textured gear teeth surface showed a significant reduction in vibration amplitude at gear mesh frequencies and minimal wear than conventional gear. It can be concluded that the researchers have kept more focus on surface coating and surface texture on the gear to significantly reduce the wear and improve meshing of gears.

5.2 BEARING

When a human invented the wheel and axle, they put a beating, made of leather or wood, and made it lubricate with animal fat. Today, bearings are used in many moving objects such as roller skates, bicycles, where two surfaces rotate or move

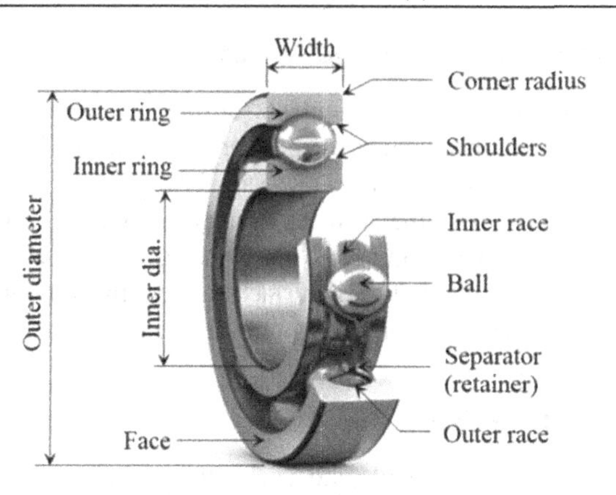

FIGURE 5.5 Nomenclature of a deep-groove ball bearing (Othman et al., 2015).

against each other due to which efficiency of product is improved. A bearing is used to reduce friction, support a load, and guide the moving parts such as wheel, shafts, and pivots. The two basic bearing categories are ball bearings and roller bearings. The grooves, where the ball rolls, are known as ball paths. However, the rollers roll on the flat surface of each race known as roller path. The separator is for holding the balls or rollers. It is positioned between the inner and outer races; the separator keeps the rolling elements evenly spaced as shown Figure 5.5.

The most popular type of ball bearing consists of a single-row of balls. In addition to the single row design, there are also double-row, angular contact, and ball thrust bearings. Roller bearing is the type of bearing where the roller can be in cylindrical, conical, spherical, or concave shapes. Roller bearing is classified into three types: tapered roller, cylindrical roller, and needle roller. The cylindrical and needle roller are non-tapered rollers. The main difference between tapered and non-tapered rollers is the shape and the curvature of the races. In a non-tapered roller bearing, the centers of each part run parallel to one another. In the tapered roller, if the imaginary lines running through the outer race and inner race, they would taper off and eventually coincide at a point even on a line extended through the bearing's center.

5.2.1 Bearing Lubrication

Lubrication in the bearing can be achieved by two methods, one is by surface coating and another is by lubricating oil. The surface coating on the bearing helps to protect the bearings against rust and corrosion and improve lubrication during working conditions. The lubricant has four major purposes in the bearing, which are discussed next.

5.2.1.1 Reduce Friction and Wear

Bearings are constantly moving elements during operation. Their moving races and rollers are rubbed against each other as well as the housings around them. Under the high speeds and heavy loads, bearings build up tremendous friction that has worn down the surfaces through rubbing action. This results in a premature failure and damage to the bearing, shaft, and housing. But when a lubricant is applied between surfaces a smooth rubbing action takes place and saves the bearing from premature failure.

5.2.1.2 Dissipate Heat

When two materials are rubbed with each other, heat is generated due to friction. This has worn the contact surface between races and shaft and housing seats. Further, this induced heat slowly damages the bearing material through deformation. In such cases, lubricant can prevent the temperature from reaching a point where it can cause severe wear. Lubricant not only carries the temperature from one place to another but also reduces the friction, resulting in less heat generation.

5.2.1.3 Protect Surfaces from Dust and Corrosion

A small amount of moisture, dirt, or dust can cause the bearing parts to corrode. Bearing needs much more clean and dust-free conditions, for smooth functioning and to operate properly. Therefore, lubrication is required that develop a protective film, which resists moisture or dust entering the bearing materials, resulting in rust protection.

5.2.1.4 Help Seals Protect Bearings

Lubricant has filled the gaps and vent present in the bearing elements. This results in the housing free from dirt or dust getting inside. A thick coating of lubricant acts as a barrier to contaminants, which assistes the seal in lubrication retention and dirt exclusion.

5.2.2 Latest Trends in Bearing Materials and Bearing Coatings

The bearing materials are traditionally made up of metallic materials, non-metallic materials, and metal–non-metallic composite. Steel is one of the most used materials in bearing industries. Due to the high coefficient of friction, significant damage to the surfaces of bearing balls occurs because of crack propagation, cavity development, and other factors, reducing the bearing life. Rathaur et al. (2019) fabricated a bearing ball of epoxy resin blended with graphite/talc micro fillers of uniform diameter of 12.7 mm and found that coefficient of friction reduced by 63% and wear resistance improved by 34% compared to the pure epoxy bearing ball under dry condition for low load application. Further, Rathaur et al. (2018) compared the 150-micron composites coating of epoxy and SU-8 resin polymer on the bearing steel. The epoxy polymer and its composite coating have exhibited good tribological properties such as a lower coefficient of friction (~0.16), high wear resistance properties (>102 times) in comparison to SU-8 and its composite. The pure SU-8 coating shows better elastic modulus and hardness than pure epoxy; however, the addition of filler enhances the properties of epoxy than SU-8. It can be concluded that polymer composite coatings can play a very important role in reducing the friction and wear at the interface and reducing the use of bulk lubrication due to which a green environment can be developed. But till now these composite coatings are developed for low load applications, but, for high load applications, researchers are still trying to get the optimum solution to reduce the bulk lubrication.

5.3 AUTOMOBILES

In automobiles, lubrication plays a vital role in all the moving parts such as gear, piston assembly, valve trains, bearings, transmissions, clutches, wiper blades, electrical contacts, etc. (Joshi et al., 2015; Priest & Taylor, 2000). It is worth noting that only 12% of the available energy in the fuel is transmitted to drive the wheels, and other being dissipated as mechanical, mainly frictional losses. The percentage of various friction losses with respect to engine components has been shown in Figure 5.6. The lubricants can be used to reduce friction, wear, and fuel consumption, which results in increased power output of the engine system. The lubricants are widely affected by loads, speed, and temperature. (Joshi et al., 2015). Surface interaction and topography in a lubricated machine element can have a significant role in the performance and durability of the components. The various modes of lubrication are required for different components and their performance. For such specific performance improvement, the additives are used in the base lubricant that helps to reduce frictional losses and leads to reduced film thickness. During in 20th century, the thickness of the lubricating film in machine elements has reduced by several orders, which are in the order of 1 mm.

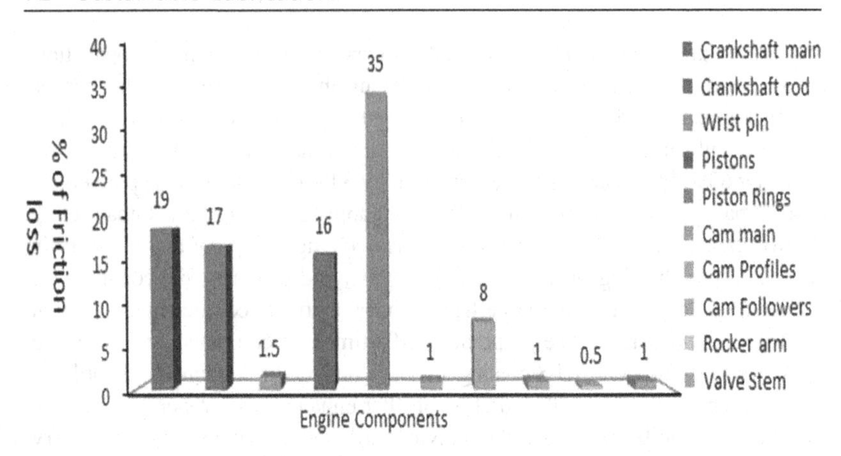

FIGURE 5.6 Losses due to friction with respect to engine components (Katiyar et al., 2019).

5.3.1 Engine Lubrication

Lubrication has played an important role in the internal combustion engine, which increased the working life of an engine. The anti-wear and extreme pressure additives are the important filler in lubricants for enhancing the life of engine. Yadav et al. (2018) demonstrated SAE 15W40 and SAE20W50 grade lubricating engine oil for wear preventive and extreme pressure properties. They conducted the test using ASTM D4172 and ASTM D2783 standards. The main function of a lubricant is to reduce friction, wear, corrosion, temperature, contamination, and shocks. The formed lubrication films protected the metal-to-metal contact of surface due to the presence of extreme pressure and anti-wear additives. Further, Ma et al. (2021) used Al_2O_3, TiO_2, SiO_2, and ZrO_2 nanoparticles as a lubricant additive in a commercial engine oil SAE 10W30 to improve the lubricity of engine oil due to which a significant resistance to wear was obtained. They dispersed 20 mg of nanoparticles in the base oil using the ultrasonication method. The results revealed that Al_2O_3 nanoparticles have reduced the wear significantly by 61%, while SiO_2 and TiO_2 showed the corresponding less reductions of 13.7% and 23.2% in the wear rate compared with pure engine oil. In contrast, ZrO_2 adversely accelerated the wear with a considerable increase of 69.69%. Among four nano lubricant additives, alumina exhibited the superior performance, up to 80% reduction of friction, and increased wear resistance by 50% compared to engine oil. These nanoparticles fill the pores on the surface and develop a thin tribofilm.

5.3.2 Latest Trends in Piston and Cylinder Lubrication

In engine, piston and cylinder wall is the main component, which is worn due to complex thermal stresses, mechanical erosion, friction, and wear. The liquid lubricant alone does not solve the problem in the long run. Therefore, researchers have suggested surface coating, which is the effective way to protect the component of engine especially the piston. Piston is the moving component in the automotive engine, which slides inside the engine chamber. The movement of the piston has prime importance in IC engines because it produces motion and generates the power for automobiles. Piston executes small translation and rotation motion within the constrained area during the reciprocating motion. Moreover, to protect the piston, piston rings are used. The main function of piston ring is to seal the piston and cylinder that is affected by mechanical, thermal, and tribological behavior of the piston. The interactions of piston rings are influenced by several parameters such as wrist-pin and crankshaft offsets, piston/liner clearance, and elastic behavior of in-contact bodies as well as thermal and mechanical distortion of liner and skirt profiles.

Further, the piston skirt is the lowest part of the piston, which has a small groove for oil retention and transporting lubricant for proper lubrication on the cylinder wall. When the piston is positioned at center, the unbalanced forces and moments perpendicular to the gudgeon pin axis act on the piston. Such action induces small translations and rotations of the piston within a defined clearance. These motions are known as piston secondary motion (Delprete & Razavykia, 2020). Furthermore, Cho et al. (2010) proposed two methods to reduce frictional losses and the amount of wear in a piston assembly. One is surface coating on the sliding surface and another is the surface roughness of skirt. They coated graphite and diamond-like carbon (DLC) on the skirt surface. A coated piston skirt exhibited smooth surface, resulting in lower friction and less wear than uncoated surface. The friction coefficent is lesser in graphite coating than the DLC, but the surface protection is lesser in graphite. Ma et al. (2018) fabricated a MoS_2-Al_2O_3 composite coating of thickness 20 micron on ZL109 substrate by microarc oxidation with electrophoresis deposition. The ceramic matrix composites coating exhibited excellent anti-wear and self-lubricating properties.

Moreover, the cylinder surface is another component where reduction of friction is very much required to improve the performance of internal combustion engine. This is the reason, He et al. (2016) demonstrated the TiO_2-reinforced Al_2O_3 coating on the cylinder wall using plasma spraying process. The coated engine was simulated and it was found that the coated material exhibited higher wear resistance and lower friction coefficient at high temperature. The coating showed a superior chemical stability to oil, resulting in less variation at increased temperature. In addition, the coated surface has shown lesser surface roughness and higher hardness. Rao et al. (2021) demonstrated the 3 mm grooved texture on cylinder liner-piston ring (CLPR) friction pairs and tested the performance of diesel engines such as the wear mass loss of the piston ring, worn surface morphologies and wear depth of the cylinder liner, exhaust gas emissions performance, service life, reliability, economy, and vibration performance of the diesel engine. The textured CLPR showed positive effect on the performance of the diesel engine and reduced the NOx emission by approximately 13.3%. Therefore, it can be concluded that the coating and surface texture on the engine component significantly improved the performance of the engine and also reduced the emission.

5.3.3 Use of Bio Lubricant in Automobile

The use of conventional lubricants and their indiscriminate disposal subsequently produce environmental problems such as pollution in water and soil, and further, the degradation of raw materials in the environment is of serious concern. Therefore, alternate lubricants in the form of synthetic lubricants such as polyalphaolefin, polyalkylene glycol, and esters were developed, and these lubricants found to be either extended the life of lubricant (drain interval) or disposed their byproducts safely after the end of service life. They are formulated by chemical synthesis rather than the refinement of existing petroleum or vegetable oils. They are generally superior from the mineral oil lubricants in terms of thermal oxidative responses, and therefore, extend their service life. Thus, the consumption of synthetic lubricants is increasing. However, the high price of synthetic lubricants impacts the market growth. Keeping this factor in view, industrial researchers have focused more on biodegradable base oil (i.e. green lubricants) which was derived from the plant oils and animal fats. The investigation on the development of bio-lubricants has received significant attention due to the fact that 50% of all lubricants worldwide will end up in the environment through spillage, usage, or improper disposal. Compared to petroleum-based lubricants, the bio-lubricants extracted from renewable origin show higher lubricity, lower volatility, higher shear stability and higher viscosity index, higher load carrying capacity. The major issue in bio-lubricant is the poor oxidation stability. Therefore, to improve the oxidation stability of bio-lubricant various chemical modifications are carried out such as transesterification, epoxidation, hydrogenation, and addition of antioxidants (Chowdary et al., 2021). Alves et al. (2013) added ZnO and CuO nanoparticles as additives to bio lubricant such as epoxidized soybean and sunflower oil. The test result showed that the improved performance of lubrication has reduced the wear and friction. Further, TiO_2 nanoparticles have been used as lubricant additive in renewable palm oil due to their outstanding chemical and physical properties. The nanoparticles are dispersed into bio-lubricant using ultrasonification method. It was observed that with increased concentration of TiO_2 nanoparticles in the base oil, the viscosity index increased by up to 4.1% (Nik Roselina et al., 2020). When, the palm oil-based TMP ester was added to the ordinary lubricant, it was observed that the addition of 3% palm oil-based TMP ester decreased the wear and reduced the coefficient of friction up to 30% in boundary lubrication regime. Further, the addition of 7% of TMP reduced the friction up to 50% in the hydrodynamic lubrication regime (Zulkifli et al., 2013). It can be concluded that that bio-lubricants have the potential to reduce the frictional losses in automobile components. The researchers are also trying to improve the performance of other bio-lubricants such as karanja oil, orange peel oil, rapseed oil, pongamia oil, mongongo oil, jatropha oil, moringa oil coconut oil, soyabean oil, sunflower oil, olive oil, peanut oil, rice barn oil. They are looking for a potentail application in automobile for making sustainable environment.

5.4 CONCLUSION

Lubrication is one of the major factors that helps to run the system in smooth and healthier way. The lubrication can be obtained by lubricating oil or by the coating on the surface of element to be protected. The lubrication in bearing and gears helps to transmit the motion with higher efficiency for a longer period without replacing the element. In automotive sector, the lubrication is very much essential not only to protect the engine but also used to reduce the soot that moves out to the environment by disposability character of the lubricant. Further, the extreme pressure and antioxidant properties of a lubricant helps the piston to produce higher expansion stroke as slide easily. Moreoever, the addition of nanoparticles as lubricant additives helps to improve the properties of lubricant. Furthermore, due to synergistic effects of the nanoparticles, the researchers are very much interested to explore more to find out the effective way to incorporate the same into the lubricating system. The less amount of nanoparticles is sufficient to improve the properties of the lubricant in higher order.

REFERENCES

Alves, S. M., Barros, B. S., Trajano, M. F., Ribeiro, K. S. B., & Moura, E. (2013). Tribological behavior of vegetable oil-based lubricants with nanoparticles of oxides in boundary lubrication conditions. *Tribology International*, *65*, 28–36. 10.1016/j.triboint.2013.03.027.

Cho, D. H., Lee, S. A., & Lee, Y. Z. (2010). The effects of surface roughness and coatings on the tribological behavior of the surfaces of a piston skirt. *Tribology Transactions*, *53*(1), 137–144. 10.1080/10402000903283276.

Chowdary, K., Kotia, A., Lakshmanan, V., Elsheikh, A. H., & Ali, M. K. A. (2021). A review of the tribological and thermophysical mechanisms of bio-lubricants based nanomaterials in automotive applications. *Journal of Molecular Liquids*, *339*, 116717. 10.1016/j.molliq.2021.116717.

Croccolo, D., De Agostinis, M., Olmi, G., & Vincenzi, N. (2020). A practical approach to gear design and lubrication: A review. *Lubricants*, *8*(9). 10.3390/LUBRICANTS8090084.

Delprete, C., & Razavykia, A. (2020). Piston dynamics, lubrication and tribological performance evaluation: A review. *International Journal of Engine Research*, *21*(5), 725–741. 10.1177/1468087418787610.

Gupta, N., Tandon, N., & Pandey, R. K. (2018). An exploration of the performance behaviors of lubricated textured and conventional spur gearsets. *Tribology International*, *128*, 376–385. 10.1016/j.triboint.2018.07.044.

He, P. F., Ma, G. Z., Wang, H. D., Yong, Q. S., Chen, S. Y., & Xu, B. S. (2016). Tribological behaviors of internal plasma sprayed TiO2-based ceramic coating on engine cylinder under lubricated conditions. *Tribology International*, *102*, 407–418. 10.1016/j.triboint.2016.06.011.

Joshi, d. s., Shah, a. v, & Gosai, d. c. (2015). Importance of tribology in internal combustion engine: A review. *International Research Journal of Engineering and Technology (IRJET)*, *2*(7), 803–809. https://www.irjet.net/archives/V2/i7/IRJET-V2I7124.pdf.

Katiyar, J. K., Bhattacharya, S., Patel, V. K., & Kumar, V. (2019). Introduction of Automotive Tribology. In: Katiyar, J., Bhattacharya, S., Patel, V., & Kumar, V. (eds) Automotive Tribology. Energy, Environment, and Sustainability. Springer, Singapore. 10.1007/978-981-15-0434-1_1.

Liu, H., Zhu, C., Zhang, Y., Wang, Z., & Song, C. (2016). Tribological evaluation of a coated spur gear pair. *Tribology International*, *99*, 117–126. 10.1016/j.triboint.2016.03.014.

Ma, C., Cheng, D., Zhu, X., Yan, Z., Fu, J., Yu, J., Liu, Z., Yu, G., & Zheng, S. (2018). Investigation of a self-lubricating coating for diesel engine pistons, as produced by combined microarc oxidation and electrophoresis. *Wear*, *394–395*(October 2017), 109–112. 10.1016/j.wear.2017.10.012.

Ma, F., Pham, S. T., Wan, S., Guo, L., Yi, G., Xia, Y., Qi, S., Zhu, H., & Ta, T. D. (2021). Evaluation of tribological performance of oxide nanoparticles in fully formulated engine oil and possible interacting mechanism at sliding contacts. *Surfaces and Interfaces*, *24*(March), 101127. 10.1016/j.surfin.2021.101127.

Nik Roselina, N. R., Mohamad, N. S., & Kasolang, S. (2020). Evaluation of TiO2 nanoparticles as viscosity modifier in palm oil bio-lubricant. *IOP Conference Series: Materials Science and Engineering*, *834*(1). 10.1088/1757-899X/834/1/012032.

Othman, M. S., Nuawi, M. Z., & Mohamed, R. (2015). Induction motor bearing fault detection using hybrid kurtosis-based method. *International Journal of Applied Engineering Research*, *10*(13), 33453–33456.

Ouyang, T., Huang, G., Chen, J., Gao, B., & Chen, N. (2019). Investigation of lubricating and dynamic performances for high-speed spur gear based on tribo-dynamic theory. *Tribology International*, *136*(February), 421–431. 10.1016/j.triboint.2019.03.009.

Priest, M., & Taylor, C. M. (2000). Automobile engine tribology - approaching the surface. *Wear*, *241*(2), 193–203. 10.1016/S0043-1648(00)00375-6.

Rao, X., Sheng, C., Guo, Z., Zhang, X., Yin, H., Xu, C., & Yuan, C. (2021). Effects of textured cylinder liner piston ring on performances of diesel engine under hot engine tests. *Renewable and Sustainable Energy Reviews*, *146*(January), 111193. 10.1016/j.rser.2021.111193.

Rathaur, A. S., Katiyar, J. K., Patel, V. K., Bhaumik, S., & Sharma, A. K. (2018). A comparative study of tribological and mechanical properties of composite polymer coatings on bearing steel. *International Journal of Surface Science and Engineering*, *12*(5/6), 379. 10.1504/ijsurfse.2018.10017968.

Rathaur, A. S., Patel, V. K., & Katiyar, J. K. (2019). Tribo-mechanical properties of graphite/talc modified polymer composite bearing balls. *Materials Research Express*, *27*(xxxx), 0–31. 10.1016/j.aiepr.2018.05.001.

Simmons, C. H., Maguire, D. E., & Phelps, N. (2020). Cams and gears. *Manual of Engineering Drawing*, 421–444. 10.1016/b978-0-12-818482-0.00031-1.

Wang, X., Zhang, Z., Men, Y., Li, X., Liang, Y., & Ren, L. (2020). Fabrication of nano-TiC functional gradient wear-resistant composite coating on 40Cr gear steel using laser cladding under starved lubrication conditions. *Optics and Laser Technology*, *126*(November 2019), 106136. 10.1016/j.optlastec.2020.106136.

Watson, H. J. (1970). Types of gears, Book: Modern Gear Production. 12–25. Elsevier, Pergamon. 10.1016/B978-0-08-015835-8.50006-7.

Yadav, G., Tiwari, S., & Jain, M. L. (2018). Tribological analysis of extreme pressure and anti-wear properties of engine lubricating oil using four ball tester. *Materials Today: Proceedings*, *5*(1), 248–253. 10.1016/j.matpr.2017.11.079.

Zhang, Z., Wang, X., Zhang, Q., Liang, Y., Ren, L., & Li, X. (2019). Fabrication of Fe-based composite coatings reinforced by TiC particles and its microstructure and wear resistance of 40Cr gear steel by low energy pulsed laser cladding. *Optics and Laser Technology*, *119*(February), 105622. 10.1016/j.optlastec.2019.105622.

Zulkifli, N. W. M., Kalam, M. A., Masjuki, H. H., Shahabuddin, M., & Yunus, R. (2013). Wear prevention characteristics of a palm oil-based TMP (trimethylolpropane) ester as an engine lubricant. *Energy*, *54*, 167–173. 10.1016/j.energy.2013.01.038.

Index

Printed in the United States
by Baker & Taylor Publisher Services